DESIGNING OUT CRIME FROM PRODUCTS AND SYSTEMS

Ronald V. Clarke

and

Graeme R. Newman
editors

Crime Prevention Studies
Volume 18

Criminal Justice Press
Monsey, NY, USA

Willan Publishing
Cullompton, Devon, UK

2005

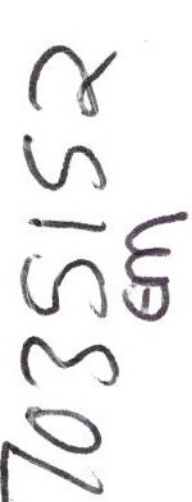

The co-editors gratefully acknowledge that this volume in the *Crime Prevention Studies* series has been sponsored in part by the Jill Dando Institute of Crime Science, University College London.

ISSN (series): 1065-7029.
ISBN (cloth): 1-881798-58-5.
ISBN (paper): 1-881798-59-3.

Cover photo (paperback edition) courtesy of the Design Against Crime Research Centre, University of the Arts, London, U.K.

Cover designs by G & H Soho, Inc.

CRIME PREVENTION STUDIES

Ronald V. Clarke, Series Editor

Crime Prevention Studies is an international book series dedicated to research on situational crime prevention and other initiatives to reduce opportunities for crime. Most volumes center on particular topics chosen by expert guest editors. The editors of each volume, in consultation with the series editor, commission the papers to be published and select peer reviewers.

* * *

Volume 1, edited by Ronald V. Clarke, 1993.

Volume 2, edited by Ronald V. Clarke, 1994.

Volume 3, edited by Ronald V. Clarke, 1994 (out of print).

Volume 4, *Crime and Place*, edited by John E. Eck and David Weisburd, 1995.

Volume 5, *The Politics and Practice of Situational Crime Prevention*, edited by Ross Homel, 1996.

Volume 6, *Preventing Mass Transit Crime*, edited by Ronald V. Clarke, 1996.

Volume 7, *Policing for Prevention: Reducing Crime, Public Intoxication and Injury*, edited by Ross Homel, 1997.

Volume 8, *Crime Mapping and Crime Prevention*, edited by David Weisburd and J. Thomas McEwen, 1997.

Volume 9, *Civil Remedies and Crime Prevention*, edited by Lorraine Green Mazerolle and Jan Roehl, 1998.

Volume 10, *Surveillance of Public Space: CCTV, Street Lighting and Crime Prevention*, edited by Kate Painter and Nick Tilley, 1999.

Volume 11, *Illegal Drug Markets: From Research to Prevention Policy*, edited by Mangai Natarajan and Mike Hough, 2000.

(continued)

Volume 12, *Repeat Victimization*, edited by Graham Farrell and Ken Pease, 2001.

Volume 13, *Analysis for Crime Prevention*, edited by Nick Tilley, 2002.

Volume 14, *Evaluation for Crime Prevention*, edited by Nick Tilley, 2002.

Volume 15, *Problem-oriented Policing: From Innovation to Mainstream*, edited by Johannes Knutsson, 2003.

Volume 16, *Theory for Practice in Situational Crime Prevention*, edited by Martha J. Smith and Derek B. Cornish, 2003.

Volume 17, *Understanding and Preventing Car Theft*, edited by Michael G. Maxfield and Ronald V. Clarke, 2004.

Volume 18, *Designing Out Crime from Products and Systems*, edited by Ronald V. Clarke and Graeme R. Newman, 2005.

Volume 19, *Situational Prevention and Child Sex Offending*, edited by Richard Wortley and Richard Smallbone, forthcoming 2005.

Volume 20, *Imagination for Crime Prevention: Essays in Honor of Ken Pease*, edited by Graham Farrell, Kate Bowers, Shane Johnson and Mike Townsley, forthcoming 2006.

Contents

Introduction 1
Ronald V. Clarke and Graeme R. Newman

1. Modifying Criminogenic Products—What Role for Government? 7
Ronald V. Clarke and Graeme R. Newman

2. Partners against Crime: The Role of the Corporate Sector in Tackling Crime 85
Jeremy Hardie and Ben Hobbs

3. Promoting Design against Crime 141
Simon Learmount

4. Breaking the Cycle: Fundamentals of Crime-proofing Design 179
Rachel Cooper, Andrew B. Wootton, Caroline L. Davey and Mike Press

5. Designing Out Crime from the U.K. Vehicle Licensing System 203
Gloria Laycock and Barry Webb

6. Security Coding of Electronic Products 231
Ronald V. Clarke and Graeme R. Newman

Introduction

by

Ronald V. Clarke
Rutgers, The State University of New Jersey

and

Graeme R. Newman
The University at Albany, New York

Criminologists have long been familiar with designing out crime from buildings, but few are familiar with the idea of designing out crime from products, even though it is not a new idea. From at least the end of the 17th century, when the edges of silver coins were milled to stop people from clipping them to make new coins, hundreds of products have been modified at manufacture to make them less readily exploited by criminals. Two criminologists with a long-standing interest in designing out crime from products are Paul Ekblom and Ken Pease, both of whom have produced important reviews (Pease, 2001; Ekblom, 2005) and who helped to make this first book on the topic possible. They each played an important part in two complementary British government initiatives to focus policy attention on the topic: the Home Office "Design against Crime" research and development program, and the Department of Trade and Industry's Foresight Crime Prevention Panel.[1] Four of the six chapters in this volume were prepared under these initiatives, but by the time the reports were complete, government interest in the subject had waned under pressure of more urgent matters, and the reports were not published. The two remaining chapters of the six (by Hardie and Hobbs of IPPR, the Institute for Public Policy Research, and Laycock and Webb of the Jill Dando Institute of Crime Science, University College London) were prepared by researchers closely in touch with the government work.

This is a lengthy and detailed book, and we have decided to keep the introduction brief by attempting to summarize only the most important conclusions of the book under the seven headings below.

1. Products Play an Important Part in Crime

A wide variety of manufactured products (though only a small proportion of all those produced) promote many different kinds of crime from theft and fraud to robbery, violence and vandalism. In general, products can serve as *tools* for crime or as *targets* for crime. Guns and spray-paint cans are tools (for violence and vandalism, respectively), while cash, cars, jewelry and VCRs are popular targets for theft. The advent of new products, such as laptop computers or ATM machines, can produce mini crime waves, or "crime harvests." Increasingly, electronic products, such as cell phones and credit cards, and the systems on which they depend for their use, are being targeted by thieves and fraudsters. But not just electronic products depend upon associated systems that can be attacked by fraudsters. For example, vehicle license documents and license plates depend on elaborate record-keeping systems for their effectiveness. These systems can be highly vulnerable to crime, as is shown in Chapter 5 on the U.K. vehicle licensing system.

2. Modifying Criminogenic Products Can Be Highly Effective

Relatively few product changes have been evaluated, but some act so directly to reduce crime opportunities that their impact is self-evident. For example, few thieves seeking cash would break into public phones that could only be operated by prepayment cards. In other cases, it would be obvious if changes had not worked. For example, there would have been a media outcry if the widespread introduction of tamper-proof seals had not stemmed further poisonings of the Tylenol variety. In fact, most of the published evaluations of product changes show positive results, sometimes quite spectacular effects. For example, as discussed in the next chapter, U.S. cell phone companies virtually eliminated "cloning," which had cost them more than $800 million in 1995, by extensive modifications made to their software systems. There was little evidence of displacement as a result of these modifications because other forms of cell phone fraud showed only modest rises when the cloning epidemic was eliminated. Despite these successes, those seeking to prevent crime through modifying products can expect to be engaged in a perpetual "arms race" with criminals, who will continually seek new product vulnerabilities. However, *not* to do so could leave opportunities wide open as existing means of prevention lose their potency.

3. Most Products Have Been Modified for Commercial Reasons

Of the many hundreds of products that have been modified at manufacture, most are *business* products modified for commercial reasons. These include: park furniture, bus shelters and railway carriages that have been modified to prevent vandalism; cell phones, cable TV boxes and parking meters that have been modified to prevent theft of service; credit cards and ATMs modified to prevent fraud; shopping carts and gas pumps modified to prevent theft; buses and taxis to prevent robberies of drivers; and banknotes, postage stamps and vehicle license documents to prevent fraud.

4. Manufacturers Have Been Reluctant to Change Products in the Public Interest

Manufacturers have resisted changing products when they are not directly harmed by the crime. For example, manufacturers have until recently resisted calls for improving the security of cars in order to reduce car thefts. Since vehicle-related thefts comprise a large proportion of all crime, this means that the crime rates of many countries have been much higher for decades than they might have been had manufacturers acted faster.

Manufacturers have been particularly reluctant to act in the public interest when: (1) they profit from the crime (for example, by the sale of replacement items for ones stolen); (2) they are scrambling to develop new products; (3) changes are costly, inconvenient or of unproven value; (4) the crimes are considered trivial and public concern is not high; and, (5) solutions are controversial (as in the case of "safer" handguns). It must also be said that many manufacturers remain unconvinced of the case for product change. Like the rest of the public, many of them believe that the best way to deal with crime is to beef-up enforcement and punishment, a view consistently reinforced by the rhetoric of politicians appealing to the popular vote. This makes it easy for businesses to argue (even when not cynically pursuing their own interests) that it is not their products and practices that need to be changed, but the police and the criminal justice system. This means that if product change is to become an established part of government crime control policy, businesses (and people in general) need to be educated about the limited capacity of the criminal justice system to deal with crime. They need also to accept some responsibility for controlling crime as is discussed in Chapter 2.

5. Design Professionals Have an Unexploited Role in Product Change

With some notable exceptions, such as the work done by Lorraine Gamman and her colleagues at the London Central St Martins College of Art and Design on *Karrysafe* bags and *Thief Proof* chairs (www.designagainstcrime.com), designers have shown little interest in designing out crime from products. This may be because they too believe that the solution to crime lies in a more effective criminal justice system. However, Chapters 3 and 4 in this volume (by Simon Learmount and by Rachel Cooper and her colleagues) suggest that designers may be open to greater involvement in design against crime, if given a lead by their professional associations and provided with appropriate design models to follow.

6. Governments Have Rarely Taken the Initiative in Promoting Product Change

To date, governments have taken a largely reactive role in product change, designed to solve specific problems connected with particular classes of products, rather than to develop a longer-term policy position. For example, in the wake of the 1982 Tylenol poisonings in Chicago, the U.S. government acted quickly to mandate standards for tamper-proof packaging. In many cases, governments have been pressed to act by the media, the police, and a variety of pressure groups, which have sometimes been motivated by crusading politicians and even researchers. In the United Kingdom, governments have generally sought to achieve change through behind-the-scenes discussions with industry, but in recent years they have shown more inclination to stir up the media and the public. For example, as mentioned in Chapter 1, a few years ago the government used the media to pressure phone companies to disable stolen cell phones in order to reduce muggings. In the United States, governments have shown greater willingness to force change through legislation and, very recently, even through litigation (more than 30 city and county governments have sued the gun industry to recover the costs of dealing with gun violence). Governments everywhere have been reluctant to offer subsidies or tax exemptions to manufacturers who make changes, or to tax or fine those who do not.

7. Governments Must Develop Research and Development Capacities in Order to Take a More Active Role in Modifying Criminogenic Products

There is a clear case for governments becoming more proactive in regard to product change. They should seek to predict and prevent "crime harvests" resulting from the introduction of new products and should find ways to exploit the crime prevention potential of biometric recognition, source tagging, smart cards and a host of other new technologies. However, governments will face many difficulties in becoming more proactive, including the reluctance of business and industry to accept their roles in causing (and preventing) crime, pressures to avoid business regulation, difficulties of obtaining proprietary information needed to make the case for product change, difficulties of obtaining international cooperation in changing products, and difficulties of limiting information about their criminal misuse in the age of the Internet. Greater than any of these difficulties, however, is the size and complexity of the undertaking resulting from the variety of industries and businesses involved, the sheer number of criminogenic products, the bewildering speed of their development, their technical nature and the complexity of the information and service-delivery systems of which many are a part. To overcome these difficulties governments must develop research and development capacities focused on designing out crime from products. It is worth remembering that all these difficulties have been sufficiently overcome in the field of product safety to make a significant difference to people's lives.

In summary, certain products increase crime by serving as tools or as attractive targets, and sometimes their introduction can cause small crime waves. Modifying products can reduce and even eliminate specific categories of crime. Manufacturers have sometimes been reluctant to act in the public interest when they are not commercially harmed by the crime. Governments therefore have the right, if not the responsibility, to take a more active role in modifying criminogenic products. In order to do this effectively they must establish small research and development capacities to promote product change and work cooperatively with business.

Address correspondence to: Ronald Clarke, School of Criminal Justice, Rutgers University, 123 Washington Street, Newark, NJ 07102 USA (e-mail: rvgclarke@aol.com).

NOTES

[1]The Foresight Crime Prevention Panel was charged by the government with looking "up to 20 years ahead, at how new technology might impact upon crime and crime prevention. As part of this it sought to consider the social changes which might occur, and how these might influence both crime and the use of technology" (Davis & Pease, 2000, p. 59).

REFERENCES

Ekblom, P. (2005). Designing products against crime. In N. Tilley (Ed.), *Handbook of crime prevention and community safety.* Cullompton, UK: Willan Publishing.

Davis, R., & Pease, K. (2000). Crime, technology and the future. *Security Journal, 13*, 59–64.

Pease, K. (2001). *Cracking crime through design.* London: Design Council.

Modifying Criminogenic Products: What Role for Government?

by

Ronald V. Clarke
Rutgers, The State University of New Jersey

and

Graeme R. Newman
The University at Albany, New York

Abstract: *Many ordinary manufactured products provide the means or the temptation to commit crime, and the introduction of new products, such as the cell phone or bank machines can create a crime "harvest." Manufacturers have modified dozens, perhaps hundreds of these criminogenic products to make them less attractive to criminals, mostly for commercial reasons, but in some cases under government pressure. This chapter reviews the international experience of modifying products, whether these are the targets or the tools of crime. It describes the range of products modified, the successes that have been achieved and the scope for further changes provided by new technology. It explains why governments have become increasingly drawn into product change, it examines the different roles they have taken, and it identifies the role of other agents of change such as the media or pressure groups. It concludes by discussing the difficulties for governments of taking a more proactive role in product change, including the reluctance of business and industry to accept*

Crime Prevention Studies, volume 18, pp. 7–83.

their roles in causing (and preventing) crime, pressures to avoid business regulation, and difficulties of obtaining international cooperation in changing products. Greater than any of these difficulties, however, is the size and complexity of the undertaking, resulting from the variety of industries and businesses involved, the sheer number of criminogenic products, the bewildering speed of their development, their technical nature and the complexity of the information and service-delivery systems of which many are a part. Governments should establish dedicated units to promote product change. These units should: (1) seek to avert crime harvests by identifying potentially troublesome new products and, (2) develop a problem-solving capacity to deal quickly with unforeseen crime threats caused by the criminal exploitation of new and existing products.

INTRODUCTION

Many everyday objects are the targets or the tools of crime, and modifying their criminogenic properties is not a new idea. To prevent people from clipping silver from the edges of coins, the coins were given milled edges from the end of the 17th century. In 1841, the Penny Red replaced the Penny Black, the world's first pre-paid postage stamp. This allowed the postal authorities to use indelible black ink to cancel stamps instead of soluble red ink, which stopped people from washing off stamps and using them again.[1] More recently, the history of the motorcar can be traced in terms of gradually improved built-in security to protect cars from theft (Ekblom, 1979; Svensson, 1982; Southall & Ekblom, 1985; Houghton, 1992; Clarke & Harris, 1992; Hazelbaker, 1997; Brown & Thomas, 2003). At first, full-time chauffeurs provided the only protection these novel machines needed. Later, as they became more widely owned and widely desired, many improvements were made in response to the growing risks of theft, beginning with door and ignition locks, and progressing to the electronic immobilisers and tracking systems built into the latest cars.

Manufacturers have made most of these crime preventive changes for commercial reasons, but they are now coming under increasing government pressure to alter their products in the public interest. This reflects the fact that in most developed countries governments are now relying less on the criminal justice system to deter crime and are placing more emphasis on the role of other public and private agencies in prevention (Garland, 1996). Wider understanding of the costs of the criminal justice system and its limits in preventing crime have prompted this shift, which has been reinforced by theories that emphasise the causal role of crime opportunities—including those provided by the multitude of manufactured products in everyday use.

This chapter reviews changes that have been made to criminogenic products at manufacture, with or without government prompting. The objective is to explore the future policy scope for this approach and to identify effective means of government intervention. The review was undertaken to complement other work commissioned by the U.K. Home Office to stimulate interest in "designing against crime" among design professionals,[2] and by the Foresight Crime Prevention Panel (see Introduction chapter) to explore new opportunities for crime and for its prevention created by technology.

The scope of the review is international and examples of product change are reviewed not just from the U.K., but also from other parts of Europe, the United States and Australia. Even if specific constitutional and legal arrangements might underlie differences among countries in their practices regarding product change, it seemed important when reviewing this new field to cull experience as widely as possible.

The review is confined to two kinds of modifications made at manufacture: (1) changes to the product itself and, (2) changes to its labelling and packaging. Labels and packages are included because these can be inseparable from the product (e.g., liquids must be in containers) or they may be intrinsic to its identity and value (brand labels).

The products covered in the review include any mass-produced items that have been altered to prevent crime. These include:

1. Products that *facilitate* crime as well products that are *targets* of crime.
2. Products owned by businesses and governments, as well as those owned by private individuals (though these uses may overlap, as in hotel room furnishings).
3. Products as small as pharmaceutical pills and as large as railway carriages; products as simple as park benches and as complex as automatic teller machines (ATMs).
4. Financial instruments, such as cheques and banknotes, and a variety of other proofs and forms of authentication.
5. Computer software and other forms of electronic information.

Not included in the review are:

1. Changes made after manufacture by product users, such as the fitting of alarms by car owners, or the retrofitting of shields to protect taxi and bus drivers from robbery and assault (though when incorporated by manufacturers in future designs these changes are included).
2. Changes made to the design and layout of buildings and facilities such as housing estates, schools and malls. These come under the well-established field of crime prevention through environmental design, or CPTED (Crowe, 1991).
3. Security improvements made to the interior or exterior of individual premises such as banks, gas stations, shops and convenience stores (though changes made by manufacturers to reduce crime associated with gas pumps, cash registers and ATMs would be included).
4. Changes made to improve the effectiveness of security products, such as safes[3] and surveillance cameras. It is assumed that manufacturers would make these changes without prompting from government in order to protect their markets. However, any changes made to prevent the misuse of security products, or reduce their collateral costs to the public (e.g., false burglar alarms), would be included.
5. Revenue changes that alter the value of products such as whiskey or cigarettes (and which consequently affect their risks of theft), or legislative changes designed to reduce access to products such as alcohol, drugs or guns.

The review begins with a summary of theory supporting product change as an instrument of crime prevention (Section 1). Section 2 describes the changes that have been made to date, covering the products involved, the crimes addressed, the nature of the changes and the preventive gains achieved. Section 3 analyses the process of change and lays the ground for a closer look taken, in Section 4, at the role that has been played to date by governments. Section 5 considers the future of government policy in regard to product change.

Some case studies of product change are collected together in the Appendix in order to assist the discussion in Section 4 of the roles taken by government, but also to provide more detail for points made elsewhere in the review. The case studies are as follows:

1. Toughened Glasses for British Pubs
2. Caller-ID for Telephones in the United States
3. Tamper-proof Packaging in the United States
4. Cheque Security in the United Kingdom
5. Smart Guns in the United States
6. The V-Chip for Televisions in the United States
7. Redesign of Banknotes Worldwide

1. THEORETICAL BACKGROUND

Until recently, most criminologists would have dismissed out of hand the idea that changing products can prevent crime. The overwhelming consensus, based on decades of research showing that a small minority of individuals is responsible for most of the crimes committed, was that only by changing the criminal dispositions of offenders could crime be effectively prevented. However, a growing body of theory and research undertaken in the past 25 years has established that offenders are as much drawn into crime by easy opportunities as they are pushed into crime by criminal propensities. Consequently, reducing opportunities will also result in preventing crime. This position has been extensively covered in other publications (for example, Felson & Clarke, 1998; Clarke, 2005), and only the main points as they relate to product change are summarised below.

1. *Crime levels in society are as much affected by the opportunities afforded by the physical and social arrangements of society (including the nature and supply of products) as by the attitudes and dispositions of the population. Since the physical and social arrangements of society are in constant flux, so are opportunities for crime.* This is the central contention of routine activity theory, first put forward to explain the rise in burglary in the United States in the 1960s and 1970s (Cohen & Felson, 1979). Two factors acting together were shown to explain this rise. The first was a substantial movement of women into the labour force, which meant that homes were left without "capable guardians" for much of the day. The second was the large increase in the ownership of TVs, audio

systems and VCRs, which provided many "suitable targets" for burglary. These two factors satisfactorily explained the rise in burglary, without the need to postulate any change in the number of "likely offenders" in society. Since this pioneering study, many other studies have shown that crime waves, large and small, can be created by the arrival of new products, including credit cards, cell phones and ATMs (Felson, 2002).

2. *Crime is heavily concentrated at particular places ("hot spots") and on particular individuals ("repeat victims"). Theft is also concentrated on particular "hot products."* These concentrations of crime are of the same order, if not greater, than the concentration of offending among a small group of the population, and their implications for crime policy are equally important. In particular, they suggest that to preserve resources preventive action should be concentrated where risks are greatest. As for hot products, the example of the motorcar has already been mentioned. Certain models are at much greater risk of theft than others, and which models are taken depends on the nature of the offence. Those taken for joyriding are quite different from those taken for resale, and both are different from models that are targeted for accessories or parts (Clarke & Harris, 1992a). The British Crime Survey provides other examples of hot products. It shows that the objects most consistently stolen from private individuals include cash, vehicle parts and accessories, clothes and purses or wallets. Annual surveys in the United States show that, while shoplifted items vary from store to store depending on stock, certain items are also consistently stolen more often. These include cigarettes and alcohol, designer apparel and training shoes, audio and video CDs and cassettes, trinkets and jewellery, and medicines and beauty products (Clarke, 1999). The acronym CRAVED summarises the attributes of hot products, which are: concealable, removable, available, valuable, enjoyable and disposable (Clarke, 1999).

3. *Crimes always serve specific purposes for the offender and, in choosing to commit a crime the offender makes a crude calculation of the costs and benefits of doing so.* This is the basic position of rational choice theory (Cornish & Clarke, 1986). It implies that the key to effective prevention is an understanding of the situational calculus made by offenders in committing specific kinds of crime. Understanding *why* joyriders take different kinds of cars from those targeted by organised criminals

stealing cars for export, and understanding *how* they set about their respective tasks, yield a host of suggestions for reducing opportunities for each of these forms of car theft. Most of the suggestions for preventing joyriding will concern the security of the vehicle itself, and perhaps the setting in which it is parked. Those for preventing theft of cars for export will also include the security of the vehicle's documentation and of the security ports and border crossings.

4. *Situational changes that increase the risk and efforts of crime and reduce the rewards, temptations and excuses will materially influence offenders' decisions about committing crime.* This is the core of situational crime prevention. More than 100 case studies have been published showing that significant declines in specific kinds of crimes have been achieved by the introduction of situational changes (Clarke, 1997; Sherman et al., 1997; Smith et al., 2002). Those involving changes to products, such as ticket machines, public phones and cell phones, will be reviewed in more detail below.

5. *Most offenders are not particularly determined and can quite easily be put off. This means that, while displacement is always possible, it is far from inevitable.* In accordance with crime pattern theory, many studies have shown that crime closely follows the routines of everyday life (Brantingham & Brantingham, 1993). For example, burglaries and car thefts tend to be concentrated along major traffic arteries, which offenders use as much as everyone else. Vandalism and a range of other petty crimes often track the routes used by children on their ways to and from school. Other studies have shown that offenders do not stray far from home to commit crimes (Wiles & Costello, 2000) and that, very often, little planning or premeditation is involved, but easy opportunities are exploited (Brantingham & Brantingham, 1993). Consequently, when these opportunities are removed offenders show limited motivation to seek new ones. This helps to explain why so little displacement has been found in most of the studies evaluating the results of situational change (Hesseling, 1994). An example from the London Underground will illustrate this point. Soon after new ticket machines were introduced in 1987, people discovered that the 50p slot could be fooled by 10p coins wrapped in silver foil. They also discovered that pressing the coin reject button returned not the 10p that had been inserted, but a 50p coin from the store in the machine. Gangs of offenders began moving from station to station to milk the machines of cash.

After fruitless attempts to arrest them, the fraud was eliminated by mechanical changes to the machines, without any substantial displacement to another form of fraud involving slugs for the £ slot. These slugs were difficult to make and required metal working facilities. Clearly, most of those previously involved in the 50p fraud were not sufficiently motivated to start making the £ slugs (Clarke et al., 1994).

6. *Product crime risks vary predictably over time with important implications for prevention.* Because of declining value, the risks of theft decline for most products as they age (unless they become antiques). Cars are an exception as their theft rates increase over their life span. Their security begins to wear out or offenders learn how to defeat it; their spare parts are in greater demand; and, as they are resold, they are kept in increasingly poor neighbourhoods, with more offenders and less secure parking. Most products, especially those defined by CRAVED, run their greatest risks of theft at retail when they are new and when they can be found in large numbers at predictable places. Not just theft, but also counterfeiting and tampering must be guarded against at retail. Most of the changes made by manufacturers to labelling and packaging are made to protect products at this stage in their lives. On a longer time scale, certain new products, such as cell phones and personal computers, are said to go through a "product life cycle," which also determines their theft risks. At first these products attract little theft because they are unfamiliar and relatively unavailable. As their popularity among consumers grows, thieves become attracted to them for personal use or for resale. Subsequently, they become widely available and relatively inexpensive, and their attractiveness for theft declines (Gould, 1969; Mansfield et al., 1974; Felson, 1997).

7. *Because of the complexity of situations giving rise to specific kinds of crime, there are usually many alternative ways to reduce opportunities for these crimes.* These alternatives have a variety of costs and benefits (Clarke, 1997). For example improved packaging and labelling can reduce the risks of tampering or theft at retail, but so can improvements made to shop security. While the ultimate costs of these improvements fall upon consumers, the immediate costs fall differently upon retailers and manufacturers. They might not be able to agree who should make the changes and might tacitly agree to absorb the cost of theft. Again, this means that consumers suffer in terms of higher prices. It also

means that taxpayers bear the costs of an increased burden on law enforcement and the criminal justice system.

8. *Because offenders are adaptable, criminals and those seeking to prevent crime are engaged in a perpetual arms race* (Ekblom, 1999; Pease, 2001). One variant of the life cycle hypothesis emphasises the role played by prevention in reducing product vulnerability in the later stages of the cycle. Thus, Home Office researchers have argued that "crime harvests" resulting from the sometimes hasty introduction of criminogenic products are diminished by retrofitting crime prevention measures (Ekblom, 1997; Pease, 1998). One example is the cell phone, which criminals in the United States quickly learned how to "clone" so that they could use copies of legitimate phones free of charge. In 1995, these frauds were costing U.S. cell phone companies about $800 million. Within four years they were virtually eliminated through a series of technological counter-measures taken by the cell phone companies, and there has been no sign of a resurgence of cloning and no new widespread forms of wireless fraud have appeared in the United States (Clarke et al., 2001).[4] Other cases where prevention seems to be winning the arms race—forging of banknotes, for example—are mentioned below, but there are also cases where the struggle between criminals and crime prevention has been more equal. For example, car theft rates are dropping in many parts of the world, but they still remain unacceptably high. Vehicle manufacturers would no doubt have made more determined efforts to improve car security if they were as much harmed by thefts as the phone companies were by cloning.

9. *Casual offenders often discover product vulnerabilities, which are then transmitted by word-of-mouth, the media and the Internet. Much larger numbers of casual offenders then exploit these vulnerabilities, as well as organised criminals who exploit them on a larger scale.* The use of slugs in the London Underground ticket machines described above seems to have followed this pattern. First, some isolated instances were discovered at a few scattered stations, and then slugs quickly began to appear at many other stations in the system. Subsequently, organised groups of offenders began systematically to milk the machines at groups of nearby stations. The same kind of a pattern held for cell phone cloning in the United States. Soon after the occasional cloned phone began to appear (presumably as the result of individual know-how), organised

criminals began to produce them on a large scale. Not only did these phones provide free phone service, but they also protected the identity of the illegitimate users. Consequently, they became attractive to drug traffickers and other criminal groups.

The theoretical propositions summarised above should make clear how certain products can produce crime waves, small or large, and how they can be modified to reduce crime. More than that, however, the propositions also provide policy makers with sound reasons for becoming involved in product change. If products can be modified to reduce the crime burdens they impose on society, governments have the right, if not the duty, to seek changes in these products. In seeking these changes, governments are helped by the facts that products wear out and must be replaced, and that manufacturers are constantly updating them for their own reasons. The question then becomes, not *whether* government should be involved in product change, but *how* limited government resources can best be used in achieving product change. To answer this question, the product changes that have been made to date need to be examined, the government role in these changes needs to be identified and the scope for future change needs to be assessed. Section 2 below begins with an examination of the changes made to date.

2. PRODUCT CHANGES MADE TO DATE

As mentioned, there is a long history of redesigning products to reduce the risks of crime. Manufacturers have made these changes to improve profits, safeguard their market positions or to exploit new sales opportunities. In making these changes, they have not been guided by crime prevention theory, but have tried merely to make things more difficult for the offender. In most cases, they have acted without government prompting and have probably given little thought to the public benefits of their actions, or to the possible costs in terms of displacement and escalation.

The review below covers the products involved, the associated crimes and the nature of the changes made. No attempt was made to provide a complete list of products changed because of the sheer scale of the undertaking. When changes to labelling and packing are included such a list would run to hundreds of items. Even then it would almost certainly be incomplete since product changes are frequently not documented and are difficult to find through literature reviews. In any case, without first classifying products in some meaningful way, the list would be difficult to interpret. However, existing product classifications based on retail or industrial sectors proved too general to be useful in the present context. Dividing products into *targets* of crime (i.e., hot products) and *facilitators* of crime was unsatisfactory because some products are both (for example, credit cards are stolen for later fraudulent use). More problematic was that most products fell into the category of facilitator, and subdividing this according to the crimes involved ran into the difficulty that some products, such as cars (see Table 1), have been altered to prevent a variety of crimes.

Another attempt to classify products, this time by the *nature of the changes* made, also proved unsuccessful. In this case, changes were classified using the then 16 opportunity-reducing techniques of situational crime prevention (subsequently expanded to 25; Cornish & Clarke, 2003), which were grouped under four main categories of increasing the *effort* or the *risks* of crime and reducing the *rewards* and the *excuses* (Clarke, 1997). Using all 16 techniques resulted in too many empty cells, while use of only the four main categories resulted in most changes falling under increasing the effort and reducing the rewards.

A third attempt, to classify products according to the *victim* (e.g., businesses, private individuals, or the public at large), was abandoned

Table 1 Changes Made at Manufacture to Cars and Crimes Prevented

Crime	Device or Redesign
Unauthorised use and joy-riding	Ignition locks; improved door locks; steering column locks; alarms; immobilisers
Theft of cars or major body parts	As above but also: parts marking; GPS (global positioning system) locators; tamper-proof licence plates; microdots[5]
Theft from car	Stronger door locks; alarms; lockable gas caps; redesigned emblems; security coded radios; removable radios; dispersed audio system
Vandalism	Retractable aerials
Assassination	Armour plating; ram bars
Illegal use of rental car	GPS locators to detect speeding (Greenman, 2001)

because so often there are multiple victims. For example, car theft harms (1) individual owners, who are caused much inconvenience even if they recover their cars; (2) insurance companies, who must make compensatory payments; (3) the public at large, who must pay larger insurance premiums; and (4) the police and the courts, whose resources are burdened (Field, 1993).

In the end, the most useful classification proved to be a simple division between *consumer products*, owned by private individuals, and *business products*, owned by financial institutions, transportation and telecommunication companies, public institutions and government bodies (when acting as service providers). Useful as it proved, this distinction is blurred by the facts that:

- Changes made to labels and packages are intended to protect consumer products at retail before they have reached the hands of consumers.
- Private individuals regard many products that have been altered (for example credit cards or banknotes) as their personal possessions when in fact they are essential tools of business.
- Some consumer products only become targets or facilitators when owned by business (for example, bathrobes are only likely to be stolen when they belong to hotels and glasses become potential weapons in pubs).

Consumer Products

The main categories of consumer products that have been changed are *cars and car parts, food and drugs (packages and labels)* and *electronic equipment* (e.g., TVs, computers and software, photocopiers, and "safer" handguns). A listing of consumer products that have been changed to prevent specific crimes is presented in Table 2.

Business Products

Many more business products than consumer products have been changed, and the main categories are:

- *Vehicles* (buses, train carriages, taxis and trucks);
- *Service-delivery devices* (public phones, parking meters, cell phones, coin fuel meters, cable-TV boxes, automated teller machines or ATMs);

Table 2 Crimes Prevented by Changes to Consumer Products

Crimes Prevented	Products Changed	Devices or Redesign
VEHICLE THEFT	Cars; motorcycles	See Table 1 for list of anti-theft devices
THEFT OF CAR PARTS	Car audio systems	Removable radios and face plates; security-coding; dispersed audio system
	Car parts	Parts marking; windscreen VIN etching; microdots
	Car emblems	Redesign of VW (Mueller, 1971) and Cadillac emblems to prevent removal by juveniles
SHOPLIFTING	CRAVED products	Large packaging for small items; source tagging
OTHER PRODUCT THEFTS	Cell phones	Account verification technology
	Computers	Automatic tracking when logged onto Internet (Evans, 2000)
	Paint tins	Dulux tamperproof tins (Design Council, 2002)
	Gas caps	Lockable caps to prevent gas siphoning
	Handbags	Alarms and strengthened material to prevent thieves cutting the bag to steal wallet in crowds (Design Council, 2002)
BURGLARY	Luggage labels	Flap to conceal address
FRAUD	Mileometers in cars	Tamper-proof design
COUNTERFEITING	Photocopy machines	In-built bank note recognition software
	Product labels	Holograms on liquor labels
	Expensive watches	Customer registration and authentication certificates

(continued)

Table 2 ***(continued)***

Crimes Prevented	Products Changed	Devices or Redesign
COPYRIGHT INFRINGEMENT	Software; music; videos; movies	Programmed to prevent illegal copying (Lake, 2001)
TAMPERING	Food and drug products	Tamper-proof packaging and seals
HACKING	Personal computers	Anti-virus software; firewalls
ILLEGAL DRUG AND ALCOHOL USE	Syringes	One-time use syringes
	Painkillers	Narcotic neutralised by chemical if time release capsules broken open[6]
	Mixer drinks and "alcopops"	Clear labelling of alcohol content and avoidance of packaging to appeal to under-18s
HARASSMENT	Phones	Caller identification devices
VIOLENCE	Cars	Bullet-proof models
	TVs	V-chips to prevent viewing of violent programs
	Handguns	Biometric recognition/electronic bracelets to identify authorised user
TERRORISM/ REGIONAL CONFLICT	Diamonds	"Fingerprinting" and authentication schemes to inhibit sales of "blood" or "conflict" diamonds, mined illegally to fund terrorism and regional wars (Rawls, 2001).

Note: Sources are cited within this table only when they are not cited elsewhere in this chapter.

- *Cash containers* (including those in many service delivery devices but also cash registers, vending machines, ticket machines, gambling machines, coin meters);
- *Furnishings and fixtures* (park furniture, pub glasses, hotel room equipment); and,

- *Financial instruments and other authorisations* (banknotes, credit cards, cheques, postage stamps, passports, vehicle registration documents and license plates).

A listing of business products that have been changed to prevent specific crimes is presented in Table 3. No doubt many products have been overlooked, including specialised metering and recording devices used throughout business and industry (scales, time clocks, taxi meters,[7] tachographs, etc). Manufacturers probably have to improve these regularly to prevent tampering,[8] but these changes rarely come to the attention of criminologists.

Effectiveness of Product Changes

Relatively few of the product changes listed in Tables 2 and 3 have been formally evaluated (defined as a quantitative assessment published in an academic journal or government research report). However, some act so directly to reduce crime opportunities that their impact is essentially self-evident. For example, few thieves seeking cash would break into public phones that could only be operated by pre-payment cards. In some other cases, it would be obvious if changes had not worked.[14] Thus, there would have been a media outcry if drop safes had failed to stem bus robberies in the United States, or tamperproof seals had not stemmed further poisonings of the Tylenol variety (*Case Study 3*). In yet other cases, such as the fitting of bulletproof shields in taxis, some basic statistical data suggest that the changes were effective.[15] Lastly, if changes made to protect businesses profits had failed to work, or worked only for while, this would not have escaped the notice of the businesses concerned, which would have taken fresh action to stem losses.

In fact, most of the evaluations that have been published (see Table 4) show positive results, sometimes spectacularly so (with pride of place being taken by the virtual elimination of cloning, which cost U.S. cell phone companies more than $800 million in 1995). However, no systematic evaluation plan guided the studies undertaken, many of which were conducted by advocates of situational prevention. The changes were often "known" to have worked and the purpose of the study was to document the evidence. All the studies were retrospective and they were often forced to use weak research designs and small numbers of cases.

Table 3 Crimes Prevented by Changes to Business Products

Crimes Prevented	Products Changed	Devices or Redesign
VEHICLE THEFT	Trucks; construction plant	Variety of built-in anti-theft devices
OTHER PRODUCT THEFTS	Shop coat hangers	Designed to reduce "rail grabbing" of multiple items
	Hotel room hangers	Clothes hangers integrated with the hanger rail
	Hotel towels, ashtrays	Monogrammed towels, ashtray, etc.
	Shopping carts	Prepayment/refund systems; redesign to reduce usefulness of trolley outside shop; electronic locking/disabling systems (Design Council, 2002; Neeley, 1998)
	Gas pumps	Self-service pumps modified to prevent pumping before payment
BURGLARY	Coin-fed fuel meters	Replaced by card-operated prepayment meters[9]
THEFT OF CASH	Public phones	Prepaid phone card (see also changes to reduce vandalism)
	Cash registers	Automatic transaction recording
	Vending machines	Strengthened cash containers. Alarms activated when tilted or switched off. Last-in-first-out system for the coin return mechanism
ROBBERY	Buses	Drop safes incorporated in buses
	Taxis	Bullet-proof shields in American taxis
THEFT OF SERVICE	Cell phones	Anti-cloning and account verification technology
	Public phones	Blocked access to long distance service and PBX systems
	Business phones	Blocked access to long distance service
	Parking meters and ticket machines	Slug detection devices

Table 3 *(continued)*

Crimes Prevented	Products Changed	Devices or Redesign
	Cable TV boxes and satellite systems	Scrambling of signals; encryption of digital signals; matching programming with consumer address (Paradise, 1995)
	Postage meters for businesses	U.S. Mechanical meters replaced by electronic digital meters difficult to corrupt (Plastiras, 1998)
	Automatic ticket gates	London Underground ticket gates are designed to prevent "turnstile" vaulting and fare evasion by checking tickets at entry and exit
FRAUD	Credit/bank cards	Account verification technology
	ATMs	In-built video-recording of customers
	Vending machines	Slug recognition devices
COUNTER-FEITING	Banknotes/credit cards/cheques	See Appendix for case studies of changes to prevent counterfeiting of banknotes and cheques.
	Passports/green cards	Many of the same changes have been made to passports, driving licenses, ID cards, stamps, etc.
	Identity cards	
	Driving licenses	
	Vehicle registration documents[10]	
	Postage stamps	
	Academic transcripts	Similar changes have been incorporated in academic transcripts by college and university printers[11]
	Tickets	Barcoding to prevent counterfeiting and fraud "Contactless" smart cards for London Underground (Benham, 2002)
	Licence plates	See Table 4 for details of changes made to Swedish number plates. The Association of Chief Police Officers has called for similar changes to be introduced in Britain (Williams, 1998)

(continued)

Table 3 ***(continued)***

Crimes Prevented	Products Changed	Devices or Redesign
HACKING	Computers	Anti-virus software; firewalls
MOTORING OFFENCES	Rental cars	Track rental cars for illegal use
	Tachographs	Automatic logs of long distance truck trips; digitisation of tachographs (Anderson, 1998)
VANDALISM	Public phones	British Telecom redesigned phone kiosks and introduced the "Oakham" booth to reduce vandalism; similar changes in Australia (see Table 4)
	Street lights	Bullet-proof shields incorporated for lights in drug-dealing areas in the United States
	Park furniture	Designed to be fixed to ground and to be resistant to deliberate damage
	Bus shelters	Open design to reduce vandalism and graffiti (Design Council, 2002)
	Glass	Toughened
	Road signs	Made from plywood in rural areas to prevent satisfying "clang" when shot at (Wise, 1982)
LOITERING	Public seating	"Bum proof" designs to discourage sleeping on benches and seats
ILLEGAL SUBSTANCE USE	Public phones	Modifications to reduce their value to drug dealers[12]
	Airliners	Smoke detectors in lavatories
	TVS and headphones for prison inmates	See-through casing to prevent concealment of drugs (LiCalzi O'Connell, 2001)
ASSAULTS	Buses	Plastic shields for bus drivers
	Railway carriages	Open designs to facilitate natural surveillance
	Pub bottles/glasses	Bottles/glasses that cannot be used as weapons
	Jail phones	Time rationing and account verification technologies

Table 3 ***(continued)***

Crimes Prevented	Products Changed	Devices or Redesign
TERRORISM/ REGIONAL CONFLICT	Airport seating	Designs that make it easy to see suspect packages left beneath the seats (Design Council, 2002)
	Car Licence plates	Modified numbering systems to prevent identification of ethnicity of car owners.[13]

Note: Sources are cited within this table only when they are not cited elsewhere in this chapter.

Even without these limitations, the studies leave some important questions unanswered. While they may show that product changes can sometime be highly effective, they do not indicate which kinds of changes work well and which do not. They also tell us little about the durability of benefits achieved through product change. How often does technological development make changes redundant (the removable car radio) and how long do modifications resist renewed criminal attack? Finally, they tell us little about the cost-effectiveness of measures and about the extent to which their crime prevention benefits are eroded by displacement.

These questions are not merely of academic interest, but are of vital concern to governments when considering their future roles in product change. To answer them, governments will have to invest much more heavily in research. They will also need to develop a new research relationship with business, because it is businesses that have made most of the product changes. Until now, governments have generally taken the position that businesses should evaluate their own crime prevention measure since they are the main beneficiaries. Unfortunately, businesses have been resistant to evaluation for a variety of reasons: the results of the measures may be obvious to them; they avoid expenditure that does not lead to increased profits; and they fear giving advantage to competitors by the release of information about effective (or ineffective) practices.[16]

However, it is clear from examples listed above that product change can bring considerable public benefits, and that failure to change products can result in significant public costs. The full extent of these costs is not always readily apparent. For example, shoplifting, assisted by easy opportunities, often supports a drug habit, while cloned phones were a boon to organised crime and drug traffickers in the United States. Likewise,

Table 4 Evaluated Product Changes

Product	Device or Redesign	Study	Findings
Cars	Steering column locks introduced in Germany (1962), the USA (1970) and England and Wales (1971)	Webb (1997)	Car theft rates decreased/stabilised after steering column locks introduced. Best results in Germany, where locks were compulsory for all cars (new and old) from 1962
Car radios	Security coding	Braga & Clarke (1994)	Theft from German-made cars began to decline in U.S. and Germany in the late 1980s following introduction of security coding
Major body parts of cars	Parts marking	Harris & Clarke (1991); Rhodes et al. (1997)	Parts marking of "high risk" cars produced, at best, modest reductions in professional auto thefts
Vehicle number plates	Laminated and tamper-proof plates with VIN mandated for Sweden in 1971	Jill Dando Institute of Crime Science (2002)	Production and distribution of plates is highly secure. Sweden has similar joyriding rate to E& W, but much less professional theft
Private phones	Caller identification made available in New Jersey in 1988	Clarke (1992)	Introduction of Caller-ID resulted in immediate 25% drop in harassing/obscene phone calls in areas with service
Cell phones	User and account verification technologies enhanced in the United States in 1990s	Clarke et al. (2001)	Losses from "cloning" were reduced from $800 million in 1995 to $50 million in 1999. Minimal displacement; highly cost-effective

Table 4 ***(continued)***

Product	Device or Redesign	Study	Findings
Public phones	Blocked access to long distance lines and PBX exchanges	Bichler & Clarke (1996)	Phone frauds of more than $1 million per month at the Manhattan bus terminal were virtually eliminated. No apparent displacement
Public phone kiosks	Coin boxes and other equipment strengthened; reinforced glass used; kiosks replaced by booths	Markus (1984); Challinger (1991); Bridgeman (1997)	Target hardening substantially reduced vandalism and theft from public phones in Australia and U.K. in 1970s and 1980s. In South Australia and the Northern Territory incidents of deliberate damage declined from nearly 6,000 in 1988 to about 1,100 in 1989
Jail phones	Time rationing and account verification technologies	La Vigne (1994)	Inmate violence connected with abuse of phone privileges reduced by 50% at Rikers Island
Credit cards	Anti-counterfeiting features in card; account verification technology	Webb (1994); Levi & Handley (1998)	These improvements together with more secure delivery of new cards led to a 40% decline in frauds losses, which fell from about £165 million in 1991 to about £97 million in 1994
Buses	Drop safes/exact fare systems in late 1960s	Stanford Res. Inst. (1970)	Robbery of bus drivers in New York and 15 other U.S. cities reduced by more than 90%

(continued)

Table 4 ***(continued)***

Product	Device or Redesign	Study	Findings
Buses	Polycarbonate protective screens for Cleveland Transit driver-operated buses	Poyner & Warne (1988)	Installation of screens on 180 buses "almost entirely eliminated" assaults on divers, previously occurring at a rate of one per month
London Underground ticket machines	Blocks on slug use	Clarke et al. (1994)	Blocks on use of 50p coins (costing £135,000) eliminated cash thefts to value of more than £250,000 per year
Parking meters	"Duncan" meter with slug-rejecter device and coin window	Decker (1972)	Introduction of "Duncan" meter led to an immediate 30–80% drop in slug use in New York City neighbourhoods

Notes:

(1) Evaluations of retrofitted drop safes in buses and bus-driver shields were included because these were subsequently adopted at manufacture.

(2) Only the most recent and/or comprehensive studies are listed in cases where more than one evaluation was undertaken.

easy theft of tempting targets like cars and their contents can be the first step for some young people on a criminal career. This gives government a strong incentive to see that evaluations are made of changes to business products. It also reinforces the general point that investment by government in private prevention could be more cost effective than increased investment in the criminal justice system (van Dijk, 1994).

Summary

This review of product changes made to date has shown that very many different products have been modified to prevent crime. Indeed, the variety suggests that almost any product could be misused for crime. While also quite various, the crimes prevented fall into several main categories, which are as follows: theft of products (including vehicle and vehicle parts); theft of services; theft of cash; product tampering; robbery and burglary; fraud; counterfeiting and illegal copying; violence; and vandalism.[17]

Most of the changes have been made to *business* products, which are essential tools for conducting business or delivering service. Of changes made to *consumer* products (i.e., ones intended for private ownership and use), most involve labelling and packing. Since many of these changes are designed to prevent crimes occurring at retail, such as shoplifting and product tampering, the principal beneficiaries once again are businesses. Apart from packages and labels, the most often-changed consumer product is the motorcar. The security of new cars is almost continuously being improved to cope with new crime threats or changes in criminal methods, and to take advantage of new technologies. These technologies are almost daily expanding the scope for product changes, which already show great variety and, sometimes, considerable ingenuity.

Relatively few evaluations have been published of product changes. These often show substantial crime reduction benefits, but they leave many questions unanswered about the generality of these benefits, about their longevity and about their cost effectiveness. A much greater investment is needed in evaluation in order to help governments think about their role in product change. This investment will also require governments to develop a new research relationship with business, which plays the dominant part in product change.

In order to prepare the ground for a closer look at the roles played by governments in product change, Section 3 of this paper will examine the processes involved in product change, the parties involved, and the ways in which they have brought pressure to bear.

3. THE PROCESS OF PRODUCT CHANGE

Most product changes are made by businesses for straightforward commercial reasons: to protect the integrity of services provided to the public; to stem cash losses; to prevent a decline in sales; to avoid product liability suits (resulting from tampering, for example); or to provide competitive advantage. To repeat some of the examples mentioned above, vending machine operators and public phone service providers have strengthened cash containers to prevent theft; bus and train companies have required vandal-resistant materials to be used for seats and other fixtures; manufacturers of expensive wrist watches have issued certificates of authenticity; distillers of high-priced alcoholic beverages have included holograms in labels; financial institutions have established automatic account verification procedures for credit cards; retailers have required small, easily shoplifted products to be packaged at manufacture in bulky containers; and self-service gas stations have installed gas pumps requiring prepayment. Governments have initiated similar changes when acting as essential service providers. For example, they have continuously sought to upgrade the security of banknotes, passports and vehicle registration documents.

Most of the changes to protect government and commercial interests have been relatively uncontroversial and have attracted little public comment. They quietly appear and become part of everyday life. Manufacturers make them on their own initiative or in agreement with business customers, including governments, retailers (and retailers' associations), financial institutions and service providers such as train and bus companies.

In some cases, business has failed to act even when it is the main victim. For instance, rather than introduce costly prevention, many retailers prefer to tolerate a certain amount of shoplifting so long as this does not significantly impact profits. (In the United States, they can even write off these losses against tax.) Unfortunately, their calculations do not include the public costs of prosecuting and sentencing apprehended shoplifters. Sometimes these failures to act can have serious consequences, as when organised criminals began to make a business of cloning phones and of counterfeiting credit cards.

Businesses have also failed to act in cases where product changes would protect the public, but would bring no special benefits for themselves. For example, vehicle manufacturers have long been criticised for doing too little to improve car security. They have claimed that the public would not pay for the increased security, but others have accused them of profiting from theft through sales of replacement cars or parts (Brill, 1982; Karmen, 1981). Only when public safety is threatened, as in the case of product tampering, do manufacturers act speedily, though, again, this is partly self-interested. Sales of pharmaceutical and food products would have been severely hit had action not been taken to prevent tampering (See *Case Study 3*).

It seems particularly difficult to get manufacturers to act in the public interest when:

- They are scrambling to develop new products, such as cell phones, before crime vulnerabilities are manifest.

- Changes are costly, inconvenient and not of proven value. British Telecom was unwilling to speed up the introduction of caller identification solely on the basis of early suggestions from the United States that this had prevented obscene phone calls.

- The crimes are considered trivial by dominant interests and public concern is not high (obscene phone calls again provide an example).
- The solutions are controversial owing to cultural factors (U.K. drinkers are said to prefer their beer served in ordinary glasses, not ones made of toughened glass) or concerns about civil liberties (caller identification and, in the United States, the right to bear arms).

In the struggle for change, a wide range of parties can become involved on either side of the debate. Apart from government in its role of public watchdog, these parties include the following:

1. *Pressure groups*—Consumer associations and motoring organisations have brought pressure to bear on manufactures to improve the security of cars. For example, the British consumer magazine *Which?* (1991) has published data showing how easy it is for thieves to break into new cars. In some cases, pressure groups have opposed product changes. For example, an alliance of women's organisations fearing that caller identification would reveal the locations of women's refuges and the American Civil Liberties Union seeking to protect the freedom of callers ensured that this service could only be offered in most U.S. states with a call blocking facility (see *Case Study 2*). Occasionally, pressure groups have taken up positions on either side of a debate, as in the case of Handgun Control and the National Rifle Association in the debate about safer guns (see *Case Study 5*).
2. *The media*—Pressure groups often make use of the media to advance their cause, but the media can also act independently. The *New York Times* has for many years advocated a variety of controls on firearms, including the production of safer handguns, and it also played an important part in publicising the high risks of robbery and homicide of immigrant taxi drivers in New York. This campaign played a part in the measures adopted to deal with the problem, including a requirement for built-in shields between driver and passengers. Quite frequently, the media also fans public concern about particular problems—phone box vandalism, robberies at bank machines and TV violence—that leads to product change. This coverage is particularly intense when a prominent person falls victim (as when Donna Shalala, Health and Human Services Secretary in the Clinton administration,

fended off robbers at an ATM in Washington in March 1999), or following some catastrophic event such as the Tylenol poisonings (*Case Study 3*) or a school shooting in the United States.

3. *The courts*—In the United States, customers who have been robbed at ATM machines or assaulted in their rooms commonly bring negligence suits against banks and hotels (Kennedy & Hupp, 1998; Guerette & Clarke, 2003). Product liability suits are much less common, though the introduction of tamperproof seals for food and pharmaceutical products was hastened by the fear of such suits (see *Case Study 3*). More than 30 city and county governments in the United States have attempted to sue gun manufacturers for irresponsible manufacturing and sales practices. These suits, which involve novel legal issues, remain unresolved (see *Case Study 5*), though *public nuisance* suits focused on irresponsible sales practices are currently making more progress than *product liability* suits focused on gun design (Butterfield, 2002).

4. *Insurance companies*—Insurance companies sometimes provide discounted car insurance policies for models with built-in theft prevention devices. In some U.S. states, they are required to do so by law. The NRMA (National Roads and Motorists Association) in Australia, which is a hybrid insurance and motoring organisation, produces annual theft data for all models of car on the road in an attempt to draw attention to the need to improve vehicle security (Hazelbaker, 1997). The Highway Loss Data Institute in the United States, which is a road safety research organisation funded by the insurance industry, goes somewhat further. It not only publishes similar league tables, but it engages in discussions with manufacturers to get them to remedy specific security weaknesses on particular models (Hazelbaker, 1997). The Association of British Insurers has a testing centre at Thatcham in which vehicles are subjected to simulated criminal attack; their performance influences the "insurance band" a given model will be assigned to, and hence the premium the owners must pay. Needless to say, the manufacturers take this process seriously.

5. *The police*—Although the police may be among the first to become aware of problems caused by particular products, they have played only a small role to date in calling for product changes. Again, the exception is cars, where police often complain publicly about poor security. In 1990, it is reported that the Major Cities Police Chiefs Association, representing 63 cities in the United States and Canada,

successfully petitioned General Motors to improve the design of steering column locks on their vehicles (Staff, 1990).

6. *Elected representatives*—Politicians quite frequently become advocates of product change. For example, Rep. Charles E. Schumer, Democrat of New York, has consistently pressed the gun manufacturers to produce safer guns. And advocates of the V-chip (see *Case Study 6*) included politicians, such as Rep. Ed Markey, Democrat of Massachusetts, who drafted clauses of the bill that was eventually enacted.

7. *Academic researchers and design professionals*—While few academics and design professionals have shown any interest in product change (Erol et al., 2002), some notable exceptions exist. For example, research programs undertaken by Levi and his associates on credit card and cheque fraud (see *Case Study 4*) have helped to reduce this problem, while Shepherd's pioneering work on injuries in pub fights has resulted in some pubs and clubs now serving beer in glasses that do not shatter when broken and that cannot be used as weapons (*Case Study 1*). The inventor of the V-chip, Tim Collings, was a Canadian electrical engineer who said he had been deeply affected by the murder in 1989 of 14 students, all women, by a Montreal gunman, who was a collector of graphically violent videos (Monaghan, 1997). Stimulated by a partnership between the Design Council and the Home Office, designers at Central St Martins College of Art and Design have produced design studies for restaurant chairs that reduce opportunities for bag theft (Ekblom, 2000).[18]

Closer study is needed of the roles played in product change by these "change agents," but the routes through which they bring pressure to bear on manufacturers are depicted in Figure 1. On the right of the diagram are the main change agents, whereas on the left are the main victims,[19] both listed in order of the degree of pressure they can bring to bear on the responsible manufacturers. The main point of the diagram is to illustrate the indirect route that pressure for product change must take when the *public* is the victim. Almost all other victims have direct and immediate access to the manufacturer concerned. For example, ATM manufacturers have to be highly responsive to the concerns of the banks, which are the main customers for their products (though many other businesses, such as convenience stores and hotels, are now installing ATMs.) When the public is the principal victim, however, this pressure must be exerted through a variety of other parties and, ultimately, must be translated into

Figure 1: Pressure for product change
(Arrows show directions of pressure).

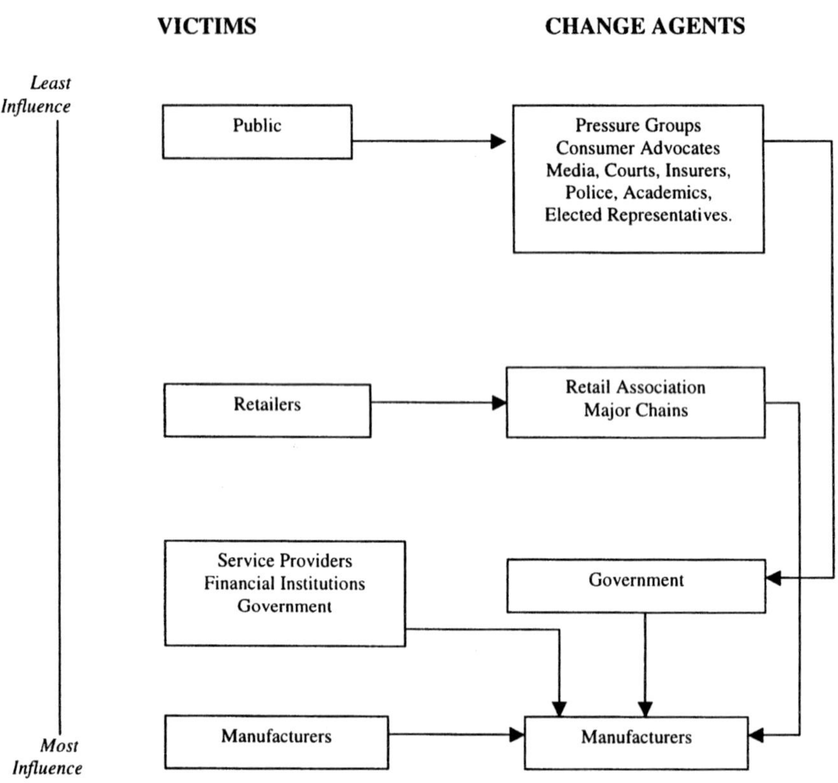

action by the government. In Section 4, a closer look is taken at the part that has been played by government in product change, before discussing (in Section 5), whether government has a greater part to play.

4. GOVERNMENT ROLES TO DATE

As explained in the next section, governments are beginning to think more proactively about product change, but until now their response to the need for modifying products has been reactive and *ad hoc,* designed to solve a specific problem connected with a particular class of products rather than to develop a longer-term policy position. Consequently, governments have adopted a wide variety of roles in product change and this variety seems to be increasing, perhaps in response to the growing possibilities for change opened up by new technology. Keeping abreast of this evolving situation is just one of the difficulties encountered when describing and classifying the role that governments have taken in product change, particularly in a review of this kind that attempts to cover all products in a variety of countries. Other difficulties are as follows:

1. The role taken by government in any specific case depends on the options available, which vary with the product, the state of current technology and with a host of other circumstances. For example, the U.S. federal government could only require V-chips to be installed in all new television receivers because the invention had recently become available and because a government-imposed system had already been developed for rating and displaying the violence content of TV programs. Absent these conditions, the government could not have acted so forcefully. Taking account of the multitude of these circumstances in a review of this kind, without becoming hopelessly enmeshed in detail, presents a difficult challenge.

2. The willingness of governments to intervene in the private sector depends on their political complexion. Thus, Conservative Party governments in the U.K. and Republican Party governments in the United States have generally been reluctant to intervene in business practices or to expand the role of government in regard to business and industry. Which varieties of governments have been in power, and for how long, can therefore have a marked impact on a country's record of product change.

3. Levels of concern about crime, which can vary considerably from one country to another or in the same country over time, can strongly influence a government's willingness to become involved in product change. For example, concern about car theft prompted the Conservative government in 1992 to publish "league tables" of the most stolen cars in an effort to get manufacturers to improve their security (Pease, 2001).

4. Whether the private sector or government offers particular services is also subject to policy. Thus, many services previously offered by government in the U.K. and some other countries—rail services, subsidized housing, water, telecommunications—are now offered through

private enterprise. In days gone by, many of these services were originally provided by business.

5. The government role in product change depends on the system of government in place in the country concerned, its constitutional powers and its relationship with the legal system. These matters are difficult enough to understand about one's own country, but they are multiplied many times over when other countries are included. Thus, most people in Britain probably have little understanding of the scope for government intervention in a federal system such as exists in the United States, which at one and the same time limits the power of central government while expanding that of state and city governments.[20]

6. The laws and legal traditions in different countries can greatly affect the scope for government intervention. In the United States, for example, there is a strong tradition, almost unknown in the U.K., of entrepreneurial lawyers filing class action liability suits on behalf of groups of citizens harmed by the products or practices of business and industry.

Despite these difficulties, the roles taken to date by government in product change can be classified into one of eight broad kinds. In order of the degree of pressure applied, these are government as bystander, arbitrator, enabler, persuader, financier, litigant, legislator and customer. The discussion of these roles is illustrated by the case studies in the Appendix.

1. *Bystander*. It was argued earlier in this review that most product changes are made quietly by industry with little involvement of other parties. Government may be barely aware of these changes, but in some cases it becomes ensnared in controversy when changes have upset the public. These upsets are often short-lived and the public becomes used to the changes, but in other cases, more might be at stake and the row might escalate. Ministers might then be drawn in willy-nilly, still essentially in the role of bystanders trying to steer a middle course. This seems to have happened in the controversy over safer bottles and glasses in pubs (see *Case Study 1*).

2. *Arbitrator*. Government has sometimes become involved in the role of "arbitrator" between industry, on the one hand, and a variety of consumer groups and pressure groups on the other. One example relates to the legal tussle in the U.S. between the phone companies and the American Civil Liberties Union (ACLU) over the introduction

of Caller-ID (see *Case Study 2*). State regulatory agencies became heavily involved in this dispute, attempting to find solutions that resolved the privacy issues, that served consumer interests and that were consistent with state laws. As the case study shows, however, technology finally came to the rescue by providing a set of choices for consumers that afforded satisfactory protection to both calling and called parties.

3. *Enabler*. Government acts in the role of enabler when it helps to bring about product changes that manufacturers desire but which require the cooperation of other parties—consumers, retailers, law enforcement and perhaps local or state governments. The best example of government as enabler lies in the regulations enacted by the U.S. Food and Drug Administration to govern tamper-proof sealing (see *Case Study 3*). These regulations were needed by business and industry to provide protection against product liability suits and, indeed, were developed with their active assistance.

4. *Persuader*. Government attempts to persuade manufacturers to change their products are of two kinds. The first, a "soft" approach, consists of behind-the-scenes discussions with manufacturers and industry representatives to persuade them of the need for change. An early example would be the agreement reached in the U.K. early in the 1970s between the Home Office and car manufacturers (represented by the Society of Motor Manufacturers and Traders) to install steering column lock on all new cars (Webb, 1997). The Home Office performed a similar role in orchestrating changes made in the early 1990s to credit cards, debit cards and cheques (see *Case Study 4*). More recently, it has adopted a more confrontational "hard" approach in regard to the security of cars and cell phones, which makes use of the media to blame intransigent manufacturers for failing to act. In both cases, it issued reports showing that the products themselves were contributing to the problem and calling attention to needed security improvements (Crime Prevention Agency, 1997; Harrington & Mayhew, 2002).

5. *Financer*. Governments can wield considerable financial power in persuading manufacturers to act. They can, in theory, let contracts for specific products meeting defined security standards; provide subsidies for product development; permit tax write-offs for expenses incurred in product development and manufacture; and impose fines on recalcitrant manufacturers. To date they have made little use of such powers

in securing modifications to criminogenic products, though a notable exception was the announcement made by Andrew Cuomo, Secretary of Housing and Urban Development in the Clinton administration, and Attorneys General from New York and Connecticut, that they would give preferential treatment in the procurement process to gun makers who adhered to a code of conduct signed by one of their number (see *Case Study 5*). This prompted a lawsuit brought by seven gun manufacturers charging that Cuomo and his co-defendants were illegally trying to influence where law enforcement agencies bought weapons.

6. *Litigant.* Handguns also provide the clearest example of governments making use of the courts to force manufacturers to change their products. As described in *Case Study 5*, in a repeat of action taken against tobacco companies, numerous city and county governments in the United States are in the process of seeking legal settlements against gun manufacturers to compensate for the costs of dealing with injuries and deaths caused by their products. Many of these suits have failed, but the final outcome is far from settled. It is clear, however, that the gun makers have been placed under great pressure by the suits to change harmful marketing practices and the unsafe design of their guns.

7. *Legislator*. Governments have most often resorted to legislation in the case of car security. Steering column locks were made compulsory in Germany for all vehicles in 1961 and in the U.S. for new cars from 1970 (Webb, 1997). Various state governments in the U.S. have required insurance companies to discount premiums for cars with in-built security devices. The U.S. Motor Vehicle Theft Law Enforcement Act 1984 required that the major body parts of new "high risk" cars be marked with the vehicle's identification number (or VIN) (Harris & Clarke, 1991), a provision that may soon be extended to all new cars. In 1998, the EU passed a law requiring all new cars to be fitted with an immobiliser,[21] and a year later Australia followed suit by issuing an Australian Design Rule with the same requirement.[22] (Western Australia went one step further by passing a law that made the installation of an immobiliser a prerequisite also for transferring ownership or re-registering of an existing car [Forbes, 2000].) Perhaps the most controversial legislation to date was that passed by the U.S. Congress in 1996, requiring all new television sets to be fitted with a V-chip to

allow parents to prevent their children from viewing violent programs (see *Case Study 6*).

8. *Customer*. For many products—passports, vehicle license plates and registration documents, banknotes (see *Case Study 7*)—government is the main customer and can require that suppliers meet certain security specifications. These specifications are constantly updated to take advantage of new technology, to accommodate new procedures (such as computerized scanning of passports), or to resist new criminal attacks. Despite its strong and direct role as customer, these changes often involve ministers and officials in much additional effort in explaining and justifying the changes to the public, and sometimes to a wider set of constituencies at home and abroad. For instance, as explained in the case study, when the U.S. government changed its banknotes, it had to take steps to reassure those hoarding its currency overseas that the old notes would still be valid.

Comment

Despite the short history of their involvement, governments have already taken a wide variety of roles in regard to product change. These have been discussed above in terms of government as bystander, arbitrator, enabler, persuader, financier, litigant, legislator and customer.

With so few examples to draw upon, it is hazardous to draw conclusions about the relative merits of these roles, though some, such as arbitrator or enabler, seem generally unproblematic. Few would also question the government's right, or indeed responsibility, to require changes to products for which it is the customer. Lastly, because most product changes are introduced by business and industry for their own reasons, it seems entirely appropriate that government's most usual role everywhere has been that of bystander—though in some cases, such as toughened glasses and bottles, one might have wished them to act more aggressively to assist change.

The more problematic roles are those assumed by governments in trying to get businesses to act when they are reluctant to do so. Some noteworthy differences have emerged between countries in their willingness to exert authority. In the United States, for example, governments have more often resorted to legislation, but sometimes with little to show as a result. As described in the *Case Study 6*, the V-chip has been a failure because parents see little need for it and rarely use it. Parts marking, which was strongly resisted by industry because of its supposed costs, seems also to have failed in its objectives because the original legislation enacted a faulty compromise: To save expense, the government agreed that only "high risk" vehicles should be marked. This was a reasonable decision in itself, but the methodology for determining "high risk" was flawed because it rested on theft statistics that failed to discriminate between joyriding and professional vehicle theft. As a result, many cars taken by joyriders were included in a program that was designed to prevent professional thefts, and some models stolen by professionals were not marked (Harris & Clarke, 1991). At best, therefore, only modest reductions in car theft could be expected from this program, which is the result found in recent evaluations (Rhodes et al., 1997). For some time, the government has been locked in disagreement with manufacturers about whether to extend the parts-marking program to cover all vehicles (the manufacturers continue to complain about the costs [Stoffer, 1997]), though the Attorney General, Janet Reno, announced on July 31, 2000, that she had determined that parts marking ought to be extended to the entire range of models sold in the United States (Padula, 2000).

The most successful legislation to date is probably that enacted in Germany requiring steering column locks to be fitted on all cars, which resulted in an immediate and sustained drop in car thefts. This was a far better result than achieved by similar legislation in the United States (which in fact made standard a feature already provided by most manufacturers) and also by the voluntary agreement reached in Britain (Clarke & Harris, 1992; Webb, 1997). However, the critical element in the success of the German legislation was not requiring manufacturers to build-in these locks, but requiring the owners of existing cars to retrofit the locks.

It is impossible to say whether government can be more successful as litigant than legislator. The only cases to review, those brought by county and city governments in the United States against the gun manufacturers, have yet to be resolved, though they have put the industry under pressure to make its guns safer. However, it is uncertain what lessons can be drawn from this legal tussle for other industries and products let alone other countries. There are so many factors that make it an unusual case, including a U.S. Constitution that permits ownership of firearms, the formidable lobbying power of the National Rifle Association, and the existence of a multitude of state, county and city jurisdictions that make such suits possible.

Guns excepted for the present, it could be argued that where persuasion has failed the case for change (or at last the particular changes sought by government) may be weak, and resorting to legal powers in the circumstances is likely to produce disappointing results. This may be why governments in the United Kingdom, aided by a less legalistic tradition and the closer contacts possible with industry in a smaller country, have tended to avoid legislation and have generally relied on persuasion to secure desired changes. In the past, they have made successful use of behind-the-scenes discussions with industry, but recently they have adopted a more public, confrontational stance, perhaps as they have come under increasing pressure to bring down crime. In the recent cases relating to vehicles and cell phones, British governments have published detailed information about security failings in order bring pressure to bear on manufacturers (Crime Prevention Agency, 1997; Harrington & Mayhew, 2002). The Home Office report documenting the considerable variation in theft risks for cars was successfully used by Home Secretary Kenneth Baker in bringing pressure to bear on manufacturers to make security improvements to their vehicles (Pease, 2001). Similarly, the Home Office report on cell phone theft caught the attention of the media[23] and, soon after its publication the five major wireless phone companies announced they would phase in a series of measures to make cell phones inoperable after theft.[24]

Whatever these differences between countries, governments everywhere have been reluctant to use their considerable financial powers (such as tax write-offs, subsidies and fines) in pursuit of change. This is surprising given that costs are the main reason that industry gives for resisting change. Considerations of the public interest, uppermost in the minds of government officials, rank much lower among the concerns of businessmen and industrialists. This reality cannot be ignored and more study should be made of the scope for government to exercise financial power. Pease has suggested that that S17 of the Crime and Disorder Act, which requires local authorities to anticipate crime consequences, could be extended to cover manufacturers. He points out that Alfred Pigou (1877–1959) was the first to advocate a tax upon industry that produces negative crime externalities and that this could include crime externalities of the kind discussed in this review (Pease, 2001). While a detailed discussion of this point falls outside the scope of the present review, we cannot forbear repeating a suggestion that one of us has previously made (Clarke, 2000), which builds upon the study undertaken by the Home Office to document the disproportionate involvement of certain cars in theft. Along the lines of the "polluter pays" principle in some environmental legislation, perhaps the manufacturers of these cars could be sent a notional bill each year (copies to the media) for the costs falling on the criminal justice system as a result of the deficiencies of their products. This would ratchet up "hard" persuasion one further notch.

5. FUTURE POLICY OPTIONS

As explained above, whenever governments have acted to bring about product change, irrespective of the particular role taken, their action has been *ad hoc* to solve a particular problem. Only recently have governments begun to think about formulating policy to guide their actions in this arena. The Australian Institute of Criminology (a government research institution) has published a brief review of the scope for "crime prevention through product design" (Lester, 2001), and the Dutch government has established a small unit, *Senter*, to study the implications of new technology for crime and crime prevention (Lester, 2001). In the United Kingdom, the crime prevention panel established by the Department of Trade and Industry under the Foresight initiative worked to this same brief and, for the past few years, the Home Office has been commissioning research (such as this review) on criminogenic products. Together with the Design Council and the Royal Society of Arts, it launched the *Design against Crime* initiative to stimulate interest in "designing out crime" among professional designers.[25]

The Foresight Crime Prevention Panel has recommended: (1) that a dedicated funding stream be established by government to focus science and technology attention on crime reduction; (2) that a national strategy be established to deal with all forms of e-crime; and (3) that ongoing programmes should be established to address crime at all stages of a product's life and to permit "horizon scanning" for future crime threats and prevention opportunities presented by new technology (Department of Trade and Industry, 2000b).

These moves are the clearest sign yet that governments are becoming more proactive in regard to product change. In this section, we consider the implications of such a policy shift and we discuss the merits of two different approaches: (1) the anticipatory design approach foreshadowed by Foresight and, (2) a problem-solving approach drawn from situational crime prevention and problem-oriented policing. But first we should mention some of the obstacles that governments will encounter in pursuing more proactive policies. These can be summarized as follows:

1. Despite the major policy shifts brought about by changed academic and official understanding of crime and its control, the public and business remain largely out of this loop. For them, the solution of crime is still to catch and punish those responsible.[26] As Garland (1996) has pointed out, this view has been consistently reinforced by the draconian rhetoric of governments seeking to win popular approval even when, at the same time, they have been pursuing more enlightened prevention policies. The prevailing public view of crime makes it easy for businesses to argue (even when not cynically pursuing their own interests) that it is not their products and practices that need to be changed, but the police and the criminal justice system.

2. Whatever their political complexion, governments are reluctant to extend their functions and are especially reluctant to intervene in business. While the threat of terrorism has supposedly made the public more accepting of government intervention, it seems likely that business will continue to be protected from intervention unless terrorism is directly involved.

3. It can be difficult for governments to obtain the proprietary information from industry that may be needed to design and implement product changes. This information might have commercial value and, in some cases, publishing it could hurt the sales of manufacturers whose products are shown to be more at risk of theft than those of their competitors. Lastly, where contemplated changes afford individual manufacturers no commercial advantage (as in the case of the V-Chip) they have little incentive to provide data.

4. Products originating in Japan and other countries with low crime rates often have less in-built protection. This is one reason why the Honda Accord and Toyota Camry (Japanese-designed cars) topped the list of most stolen cars for many years in the United States. Governments face a particular problem of negotiating with these overseas manufacturers.

5. EU regulations prevent any single member country from mandating security standards for all products of a particular class, whether domes-

tic or imported. Obtaining EU-level agreement on such regulations, particularly when crime risks vary considerably among countries, can also be difficult.

6. Legislation designed to promote free markets make it difficult to prevent sales of legitimate products that criminals use as tools. These include magnetic strip decoders/encoders (used by criminals to produce fake credit and debit cards), wireless scanners (for capturing telephone numbers needed for cloning cell phones[27]) and devices to reprogram PINs for car radios (needed if stolen radios are to be used or sold).
7. Freedom of information laws and the Internet make it easy for anti-business interests and other groups to spread information about methods of committing product-related crimes such as credit card fraud, telephone fraud and counterfeiting.

Problem Solving Versus Anticipation

Greater than any of the above obstacles faced by government in pursuing a more proactive policy of product change is that presented by the size and complexity of the undertaking. This is due to: the variety of industries and businesses involved; the sheer number of criminogenic products; the bewildering speed of their development; their technical nature; and the complexity of the information and service-delivery systems of which many are a part. Together this suggests that uniform solutions will be of limited application and, if so, preemptive action to nip crime harvests in the bud, though prudent and necessary, may not be enough on its own. It could even lead to much wasted effort through focusing on products that have become outdated and are no longer of interest to criminals, or repeating solutions that criminals have already learned how to defeat.

This is not an argument for neglecting past experience. Effective crime prevention, like any other body of expertise, relies on the systematic accumulation of knowledge and practice. Translating this knowledge into practical guidelines and incentives to prevent the emergence of future problems is both important and necessary. It is also important to invest in the building of designers' capacity to develop new crime-resistant products.[28] But in a world that is changing at ever-increasing rate, there is also a need to respond quickly to new and unforeseen crime threats by developing tailor-made solutions. Here, the difficulty lies not in identifying troublesome products early enough, since the police and media seem well attuned to these crime waves. Rather, it will be in discovering quickly enough just how the product is exploited, by whom and under what conditions, and with what resources (Ekblom & Tilley, 2000). Only with detailed knowledge of these matters is it possible to design and implement effective solutions.

A well-tried methodology for this kind of problem solving work is provided by the action-research model employed in situational crime prevention and problem-oriented policing. This model consists of several sequential stages: identifying specific problems; analysing the conditions giving rise to these problems; identifying a range of possible solutions; assisting the implementation of the most acceptable and promising of these; evaluating the results; and starting all over again if these results are disappointing (Ekblom, 1986).

Dealing with actual current problems rather than hypothetical future ones would lend greater urgency to government efforts, while the evaluation stage built into the action research model would help to focus these efforts on the most realistic solutions. A problem-solving approach would therefore help the government to use its limited resources to greatest effect—or to "get the grease to the squeak"—which for many years has been a watchword of crime prevention practice (Hough & Tilley, 1998).

A Balanced Approach

We have suggested above that governments are poised to adopt a more proactive policy of product change, and we have argued that this policy should be composed of two complementary components: (1) an anticipatory component designed to forestall crime harvests; and (2) a problem-solving, retrofit component to deal quickly with new, unanticipated problems. The correct balance to be struck between these two components will only emerge through experience. Even if more hope for the future is placed on anticipatory action, problem solving should be given equal weight to begin with since solutions are urgently needed for problems caused by some existing products (see Table 5).

Research and Information Needs

If governments are to become more proactive in seeking product change, they will also need to make a greater investment in publicly promoting the concept, in developing a body of relevant expertise and in commissioning research. We endorse previous suggestions that this investment should be

Table 5 Criminogenic Products for Which Redesign Is Needed

Product	Redesign Needed
Cars	Ignition interlock and keypad ignition systems to prevent drunk driving[29] Continuous electronic ID monitoring to prevent theft Electronic control of speed on highways
Trains, buses, trucks	Ignition interlock and keypad ignition systems to prevent drunk driving
Airliners	Breathalyser with video-recording built into start up procedures CCTV surveillance of cabin Strengthened/sealed cockpit door to prevent terrorism Computer-mediated facial recognition of legitimate crew
Automatic teller machines	Biometric recognition techniques
Fertilisers and chemicals	Remove ingredients that facilitate bomb construction
Cordless drills and other hand tools	Reduce their value for breaking into premises, vehicles and vending machines
Lock tools and designs	Breaking and entering, safe cracking, password and encryption breaking for data theft and extortion
Information: Maps, building plans, bomb recipes	Restricted access to prevent terrorism, breaking and entering
Digital cameras	These assist child pornography because pictures do not need professional processing
Phones	Ways needed to prevent illegal phone tapping and other invasions of privacy
Cell phones	Disable after theft
Phone kiosks, bus shelters and railway carriages	Scratch-resistant and shatter-proof glass

Table 5 *(continued)*

Product	Redesign Needed
Magnetic stripe gift cards	Prevent theft of stolen numbers by hackers and counterfeiters (Sullivan, 2001)
Spray paint cans	Reduce their usefulness for graffiti[30]
Pharmaceuticals with alcohol or narcotics	Replacement of addictive/intoxicating ingredients in mouthwash, decongestants and pain relievers

in the form of a research and development unit wholly dedicated to these tasks (Pease, 1997; Clarke, 1999).

Promoting the concept of product change would require the unit to be fully conversant with the results of "horizon scanning"; to keep abreast of overseas developments and those occurring in product safety, road accident prevention and other related areas of work (Christie, 2000); to make presentations to professional audiences (including industrialists, business leaders, government officials, academics and design professionals); and to use the media to inform the public of the benefits to be achieved. In order to develop bodies of expertise, the unit would need to develop consultancy relationships with a wide range of individuals in industry, business, the police and the universities, and would need to establish consultative committees composed of representatives from these fields. The research it would need to commission would span a wide range of scientific, technological and criminological topics. This review alone has identified several topics deserving of more research, including: the extent to which criminogenic products are exploited by organised crime and by criminals who are supporting drug and alcohol habits; the roles of pressure groups and other key actors in securing or inhibiting product change; the scope for governments to exercise financial powers and incentives in respect to product change; the research relationship between business and government; and the cost-effectiveness of product changes, including studies of displacement.

The Foresight Crime Prevention Panel (Davis & Pease, 2001) and the Home Office have undertaken some of these promotional, development and research tasks, but have not done so recently. In any case, assigning these tasks to a designated unit would raise their priority and their visibility to business and industry. If the unit were permanent, this would also greatly assist the development of relevant body of expertise. *Senter*, the unit established by the Dutch government was placed within the Ministry of Justice,[31] but whether this is the best arrangement, and what is the best balance between commissioned and in-house work, are technical questions lying beyond the scope of this review.

CONCLUSIONS

There can be no end to the criminal misuse of legal products. Even if dozens, perhaps hundreds of products have been successfully altered to prevent crime, criminals will continue to seek new ways of exploiting products so long as the rewards are great enough. In some cases, they will be helped by changes made by manufacturers to expand the uses of their products or improve their convenience. In addition, completely new products and associated services are being continually introduced for business and consumer use, all of which provide new opportunities for criminals. There is no end in sight to this technological development, and there seems no possibility of our economy ever being able to deliver (or indeed wanting to deliver) these products and services at a price everyone could afford. Even in such a utopia, normal human greed and vice would continue to result in their criminal exploitation.

This is not meant as a counsel of despair because, at the same time, technology is continually delivering new ways to prevent crime. Those on the immediate horizon include: biometric recognition technology and "smart" cards to reduce the opportunities for fraud; "chipping" of goods[32] or source tagging at manufacture to reduce the scope for shoplifting and other theft (DiLonardo, 1997); and use of the Internet, GPS and wireless technology to provide instant checks on ownership for a wide variety of goods such as cars, computers and television sets (Clark, 2001).[33]

As repeatedly mentioned in this review, businesses are the main victims of crime caused by products, and businesses usually provide the solutions. It is when the public is the victim, or when heavy costs fall on the public sector, that governments have been forced to intervene. It might be tempting for them to continue in this primarily reactive mode, but we believe this would be a mistake. There is ample evidence from the evaluations reviewed above that crime harvests caused by particular products are stoppable, even if they are not always predictable. The reductions achieved can be quite considerable, with many resulting benefits: police resources are conserved; the profits of organised crime are reduced; habitual offenders find it harder to support drug and alcohol habits through theft; and the public is saved much inconvenience and misery. Government could help to deliver these benefits with a relatively small commitment of its resources.

APPENDIX

Case Study 1: Toughened Glasses for British Pubs

We all know that broken glass can be dangerous, but it is difficult to use as a weapon unless there is a way to hold it without being cut. The design of beer glasses and bottles therefore, is of considerable importance in making them useful as weapons. For example, the traditional beer glass of heavy base and slender trunk, rising to a wider mouth at the top, is ideal for use as a weapon. The hand can grasp the base easily, and if need be, it is easy to smash the glass first on the bar.[34] The same applies to the older quart size bottles of beer (still available in some areas), whose top rises into a slender tapered end, which may be easily grasped and smashed against a hard surface to make a jagged-edged weapon, or may be smashed directly against the victim's face.

Some changes have occurred in the design of bottles, including the introduction of "glass cans" that do not have a slender top to grip and metal cans to replace the larger bottles in Australia. However, change in the design of glasses has been slower in coming because publicans and drinkers alike hold various beliefs about the shape of a beer glass (and its other attributes, such as its surface, temperature, how it has been cleaned, etc.) and its effects on the taste of the beer itself. The alternative possibility therefore, is to change the properties of the material used to make glasses.

Working with colleagues, J. P. Shepherd has shown that the type of glass used in traditional beer glasses can cause serious injuries (Shepherd et al., 1990a; Shepherd et al., 1990b). The glasses are made of annealed glass which, when broken, produces many sharp edges that are can be lethal, or cause disfiguring injuries. Shepherd has also shown that most glasses are smashed into the victim's face, where the glass breaks on impact (Shepherd et al., 1990a), and he has argued that a simple step to reduce the severity of injuries would be to manufacture pub glasses out of tempered rather than annealed glass. Tempered glass is the same as that used in car windscreens. It is much stronger than annealed glass in impact resistance and when broken shatters into small "sugar lump" fragments that are far less injurious. While no studies have investigated the effectiveness of toughened beer glasses in reducing injury, there is a long and proven safety record of the use of tempered glass in automobiles. Tempered glassware is no more expensive to produce than annealed glassware and it lasts some 20 times longer than annealed glasses.

In 1997, the then shadow U.K. Home Secretary Jack Straw drew attention in the course of pre-election campaigning to the problem of pub violence. He said that over half a million individuals were assaulted by others who were intoxicated and that more than 5,000 people in Britain were maimed every year as a result of attacks involving broken glasses. Straw pointed out that figures from 22 police force areas showed that the number of disorder incidents due to alcohol had risen by more than 20% over two years. The 1996 British Crime Survey had also suggested that nearly 20% of violent incidents occur in and around pubs, amounting to some 13,000 incidents a week. He called for a "three-pronged" approach to the problem, including the wider use of toughened glasses (Mason & Little, 1997).

During the course of 1997 and 1998, there were a number of meetings on drinking and violence in which various interest groups participated. These included the Addiction Forum, Alcohol and Health Research Group (Scotland), the Portman Group (an industry association), and the Addiction Research Group in Canada. These meetings generally endorsed the need for toughened beer glasses (Notarangelo, 1997; Staff, 1997). In February 1998, Mr. Straw, who was then Home Secretary, attended a Crime Concern conference and noted, "Partnership is the key to reclaiming the social and commercial hearts of our communities from the drunken yobs" (Staff, 1998a). The Crime and Disorder Bill (subsequently the Crime and Disorder Act), which he announced, would in fact require such partnerships. While it also revived some draconian police powers, this legislation recognized the need for police to work hand in hand with business and bar owners, and especially local governments, which are intimately involved with the local bars through licensing, zoning and other matters.

Mr. Straw noted in his remarks that industry had agreed to make wider use of toughened glass in pubs and clubs. In fact, it seems that the big brewers controlling most pubs have been much slower to adopt toughened glasses than independent pubs and bars (Design Council, 2002).

There are also regional variations in the use of toughened glasses. For example, they are more widely used in the North West than in London or the South East, perhaps due to the *SafeGlass-SafeCity* campaign mounted by the Greater Manchester Police and the Manchester Evening News (Design Council, 2002). These facts suggest that a stronger push is needed from government to promote the use of toughened glasses.

Case Study 2: Caller-ID for Telephones in the United States

Caller-ID is a service provided to telephone subscribers in the United States (and under other names in many other countries) that allows the receiver of a phone call to see the number of the caller, along with other personal information such as name and address from which the call is made. It is a relatively simple technology (Slawson, 1997) that, when first introduced required the subscriber to install a special device that displayed the information. The service was first offered in 1988 to residents of New Jersey. At that time, the display device cost around $70, and New Jersey Bell charged the subscriber $6.50 per month for the service. Today, most telephones of average price have the Caller-ID feature built in, and the cost to subscribers is minimal, due to the widespread competition among the different providers.

The new service was made possible by investment in new high-speed network signalling systems (Slawson, 1997) that became the standard switching technology used for the provision of Caller-ID and other services. Late in 1987, New Jersey Bell conducted a small trial offering and concluded that Caller-ID was a service that customers really wanted (Clarke, 1992). The service was quickly introduced, touting many benefits for businesses and individuals, which may be summarised thus:

- *For businesses,* Caller-ID makes it possible to: recognise incoming calls and route them to the appropriate operator; ascertain customer addresses (useful for services such as food delivery or vehicle breakdown service); and match calls to databases of credit records, spending patterns, etc. (Australian Communications Authority, 1998).
- *For individuals,* Caller-ID provides enhanced control over incoming calls: people like to know who is calling, to screen calls and return them at their leisure, to record missed calls and to enjoy a range of additional services that are often packaged with Caller-ID, such as call waiting and call trace.

An additional benefit, the focus of the present discussion, is that of reducing the level of obscene and harassing calls. These are not only upsetting to the victims, but they entail considerable costs for the phone companies and police who must deal with complaints. An early evaluation of Caller-ID concluded that the number of complaints about annoyance calls (15% of which were obscene calls) declined by 25% after Caller-ID was introduced. The study further reported that there was little evidence that annoyance calls were displaced to other parts of New Jersey that did not have Caller-ID (Clarke, 1992).

Despite its various benefits, opposition to Caller-ID quickly emerged (Hall, 1990). New Jersey Bell and other companies that followed its lead were criticized for introducing the service too rapidly without consideration for consumer concerns for privacy (Australian Communications Authority, 1998), though it must be said that subsequent surveys of consumers have revealed that consumers want it both ways: they want to control what information about themselves goes out, but they also want to obtain maximum information about calls and callers that come in. The debate concerning privacy therefore is essentially one of trying to balance these two competing demands of individual subscribers.

Some of the privacy concerns were of a specific nature. Subscribers with unlisted numbers feared that this benefit would be undermined by Caller-ID. Women's groups argued that it would enable abusers to track down partners who had taken refuge in shelters. Some law enforcement officials said it would compromise the safety of covert operators. Social workers feared it would inhibit the flow of information to crisis hotlines (Hall, 1990). Consumer groups were concerned that the associated databases of information about subscribers would be sold to telemarketers. Finally, some state officials complained that the service violated the terms of their wiretapping legislation. (For example, Pennsylvania found that Caller-ID violated its Wiretapping and Electronic Surveillance Control Act.[35])

The American Civil Liberties Union (ACLU) quickly marshalled these various concerns in a series of state court actions taken against individual telephone companies when they sought to introduce Caller-ID.[36] Many of these early cases focussed on the right of customers to block display of their numbers when making calls. This facility could be provided under the existing technology, but the phone companies argued that the costs of doing so were too great. They also argued that this facility would destroy the ability of Caller-ID to prevent annoying calls. However, the pattern was soon set by a ruling made by the Pennsylvania Public Utility Commission in November 1989, upheld in the courts, that Caller-ID must be made available in Pennsylvania with a call-blocking facility for "at risk" customers, such as those in battered women's refuges or in some law enforcement positions. A series of similar judgments were made in other states and the phone companies soon abandoned their opposition to the call-blocking facility.

A second focus of debate concerned the ability of consumers to opt out altogether of the Caller-ID service. Some telephone companies had simply introduced Caller-ID without taking steps to inform subscribers or customers what the service involved. Because the service works much better with as many customers participating as possible, the strong tendency was to automatically include an individual's number in the sending part of Caller-ID, and *charge* only those who wanted the receiving service—that is to see the number of the person who had called. This was called the "opt-out" model. Consumers had to take steps to opt-out of any part of Caller-ID or they stayed in. The alternative was to invite customers to "opt-in" to Caller-ID, which was the option favoured by the critics of Caller-ID.

While regulatory agencies and courts argued over privacy and other issues, consumers and business moved ahead and solved most of the problems raised by Caller-ID, driven by the sheer force of technological development. Although solutions vary considerably in detail from state to state, the following is probably a reasonable rule of thumb in terms of the provision of services:

1. Most plans are offered on an opt-out basis.
2. Per call blocking is offered free, and some areas also offer line blocking free, especially for unlisted subscribers.
3. The ability to block reception of calls with the caller's number blocked is widely offered.
4. 911 and other important community services can override blocking.
5. Many states have legislated the set up of "no telephone solicitation" lists, and require the bonding and registration of telephone solicitors

> (Cerasale, 1999). The European Union has also introduced similar regulations and mandated customer education upon introduction of the new service.[37]

There are also at least six other ways in addition to Caller-ID to deal with annoying telephone calls. These include call trace, call trap, call screen, selective call acceptance, call return, and various special devices. Some of these services are usually offered free, such as call trace or call trap, while others are offered on a per fee basis (Privacy Rights Clearinghouse, 2000). The convergence between computers, the Internet and telecommunications has also made possible a variety of new ways to enhance and capture the telephone numbers and personal information of telephone users (Carroll, 1999).

Case Study 3: Tamper-Proof Packaging in the United States

In 1982, seven people died in Chicago as a result of swallowing cyanide-tainted Tylenol. That event caused widespread fear about the safety of personal products that were displayed on open shelves in stores. Since that time, there have been periodic outbreaks of product tampering. In 1988, the U.S. Food and Drug Administration (FDA) received 488 tampering complaints, most of which concerned food and beverage products. Nor were these incidents confined to the United States. In 1989 in the United Kingdom, Heinz suffered an incident of tampering with its baby food products, which cost at least £2 million in advertising to relaunch the product after it was protected with tamper-resistant packaging. And in 1994, Safeway's labelled tonic water was contaminated with deadly nightshade. Although no one was hurt, the cost for recall was at least £44,000.

The Tylenol event was catastrophic not only for those poisoned, but also for the businesses that suffered enormous losses through public avoidance of their products. It took Tylenol several years to regain the 35% of market share it had in 1982, and cost the company $100 million dollars just to send out consumer warnings (Teresko, 1983). This event, together with the widespread public concern, spurred both government and business to cooperative action. The FDA moved with unprecedented speed to develop regulations that required tamper-resistant packaging for selected cosmetic products, medical devices and many over-the-counter

(OTC) drugs. It did not go further, however, and include food containers, mainly because of the much greater diversity of food products and packages.

Following intensive consultations with industry leaders and consumer groups, including Ralph Nader (Robinson, 1986), the FDA regulations were published in 1983 in a record 35 days (Hyman, 1983) from their inception. The regulations were as follows (Foley, 1983):

1. Directions for opening the tamper resistant package should be in large print, in contrasting colours and placed on both sides of the product.
2. Tamper-resistant packaging should be clearly visible and should be manufactured in a colour and design scheme that is distinct from the rest of the container.
3. Component parts of the tamper-resistant packaging that need to be manipulated to break the barrier should be clearly visible and easy to grasp.
4. There should be standards for minimum break strength so that persons with physical disabilities are able to open the packages.
5. There should be a uniformity of tamper-resistant directions and mechanisms so that the ability to open one type of tamper-resistant packaging increases the likelihood of being able to open other packaging of the same type.

The two guiding principles of these regulations were that: (1) if the seal is broken, the break must be highly visible, and (2) use of the packaging should be convenient for the consumer.

FDA inspectors were required to check on compliance, although the FDA insisted that it was more interested in achieving voluntary action. In fact, businesses had much less to fear from the FDA than from product liability suits (Hyman, 1983). If nothing else, the Tylenol incident had made it clear that manufacturers could foresee that many of their products were vulnerable to attack. This brought an element of competition into the scramble to introduce tamper-resistant packaging, not only to allay customer concerns about their products, but also for legal protection, since a legal rule prevails in product liability cases in regard to how many of a company's competitors had taken precautions. In short, there arose a strong impetus, both public and private (Teresko, 1983) to introduce tamper-resistant packaging.

The FDA regulations have been updated from time to time, with an extensive set of guidelines published in 1999 (Food and Drug Administration, 1992, 1999). These generally reflect the five rules listed above, although they do extend packaging concerns to include rules about child-resistant packaging as well. Businesses have installed tamper-resistent devices on an ever-increasing range of products marketed for personal use, partly in response to new tampering scares. Recent innovations are devices that make use of optic fibres that stop shining once the seal is broken, and many other high-tech devices (Keck, 1993) for a variety of products (Dodd, 1998). Manufacturers have pursued this course in the face of research in focus groups suggesting that most consumers rate product safety last as an issue of importance, after (in order) quality, nutrition and ingredients, convenience and price (Stillwell, 1989).

While initially there were concerns about the high cost of introducing tamper-proof packaging, these have abated, with estimates running to only about one cent a container, compared to the initial estimates of ten cents (Staff, 1982, 1989). In any case, it was recognized very early that losses could be far greater to a company in terms of reputation and care for their customer. As a SmithKline Beecham executive said, "The company's only concern is the well being and safety of its customers" (Murphy, 1991).

The Tylenol case and other incidents of product tampering taught business not only that it had a responsibility to prevent tampering, but also to think of products in much broader terms to include brand name and reputation: in effect, product integrity. Industry is now moving to design product packaging to accomplish wider security goals than simply preventing tampering by the public. There are four main types of intrusion to guard against, all of which can be defeated by using innovative technology in a single packaging system:

1. Terrorism or random attacks such as the Tylenol case.
2. Pilfering or damaging of smaller items enclosed in packaging.
3. Tampering during the manufacture.
4. Counterfeiting by using technology such as attachment of holographic or other unique identifying labels.[38]

This is a case in which a catastrophic event, followed by a series of copycat incidents (some 300), spawned widespread public, governmental and business concern. Each of these parties had a strong interest in bringing about changes that would prevent the future occurrence of this tragedy

(Hilts, 1982). This brought about an unusual level of cooperation in effecting facilitative legislation and eagerness on the part of business to produce and demonstrate tamper-resistant packaging that reassured their customers. The result has been the continued spread of tamper-resistant packaging across countries and product types (from OTC drugs, cosmetics, beverages, frozen foods [Staff, 1983], toys, multimedia,[39] credit cards[40] and even original works of art [Gialamas, 1997]) over the 20 years since the Tylenol event. In sum, all parties have benefited, and of course, an additional beneficiary has been the packaging industry, which has experienced a boom in the past 15 years.

Case Study 4: Cheque Security in the United Kingdom

According to the Cheque and Credit Clearing Company,[41] the number of cheques used for payment in the United Kingdom reached an annual total of 3.5 billion in 1990, but it is predicted that this number will decrease by about 40% by 2009 because of the rapidly expanding use of credit and bankcards. Compared to credit card fraud, cheque fraud in the United Kingdom accounts for a much smaller portion of bank services fraud, though many times more than losses to banks from simple robbery.[42] Moreover, the amount lost to cheque fraud has decreased dramatically, even while the amounts lost to credit card and debit card fraud have increased considerably in recent years. Thus, cheque fraud losses against turnover were .145% of all transactions in 2000 compared to .33% in 1991.

In 1991, an important research report (Levi et al., 1991) prepared for the Home Office described the situation of bank services fraud as one of a "stand-off" between police and business, each waiting or expecting the other to solve the problem. The report brought to public notice the extent of bank services fraud and helped to build a coalition of interests between the interested parties, which resulted in many changes to the design and manufacture of cheques, and in their delivery and processing, which brought about the reduction in cheque fraud. By 1995, major initiatives were coming into play: the introduction of "hot cheque" files, the more widespread sharing of information concerning counterfeit cheques and suspect bank accounts, and the formation of cheque fraud squads by the police which began to work in concert with banks (Levi & Handley, 1998).

Advanced anti-counterfeiting technology also began to be introduced into cheque printing around this time, which included the following features:

1. "VOID" messages that showed when a colour photocopier was used.[43]
2. Rainbow graduated colours (pantograph), which are difficult for photocopiers to reproduce.
3. Signature area printed on different pantograph colours.
4. Micro printing, which prints words so small they appear as lines or borders on the cheque.
5. Watermark certification of various kinds that cannot be photocopied.
6. Fluorescent fibres, especially yellow fibres on the back that will intensify under black light.
7. Authentication warning endorsement feature to alert the receiver to the security features of the cheque (Goldsec, 1999).
8. "Cut-and-paste" prevention background that will not match up if someone tears or cuts the cheque then tries to put it back together (Staff, 1994).

It is generally believed that most security features have about a three-year life span, after which they must be upgraded in order to keep one step ahead of the criminals (Cole, 2001). However in the case of cheques, it is likely that these high technology security features have worked more effectively and for a longer time, partly because organised crime has concentrated on the counterfeiting of credit cards over the same period. This may have as much to with the following facts as with the inherent security of cheques:

1. PIN numbers were required for the use of cheque and bankcards (but not credit cards).
2. The concerted efforts made in 1993/4 to create databases that could be checked each time a credit card or cheque cashing card was used at POS (point of sales) worked better for cheque fraud than credit card fraud because merchants generally must shoulder more of the loss from cheque fraud (Levi et al., 1991); this creates a need for the merchant to make a special effort to show "due diligence" in applying proper security procedures to ensure against cheque fraud (Stephens, 1998).
3. There has been a more concerted effort on the part of police-bank-merchant cooperation in regard to cheque fraud than there has in credit card fraud (Levi & Handley, 1998).

It is reasonable to conclude, therefore, that cheque fraud prevention technology and intervention in the service delivery system have been quite successful. The different interests can be seen at work if the series of identifiable steps through which change occurred are traced:

1. *Identifying the problem as serious.* One would expect that, once a problem is perceived as serious it would naturally lead to action to solve it, but this depends on the players and their interests. In this case, cheque printers work hard at convincing their potential customers that the problems related to cheque fraud for business are serious, so that the investment in high quality security printing is highly cost-effective. But businesses in this case were slow to accept that the problem was sufficiently serious to act upon, since the fraudulent portion of turnover was very small, and in any case there was a perception that fraud was something for the police to deal with. However, police forces had limited capacity to investigate cheque frauds, and these resources were organized in a in a hit-or-miss way, without coordination with the banks, to respond mainly to calls from merchants. Furthermore, police do not generally regard cheque fraud as a serious problem compared to other traditional crime. Finally, consumers did not perceive cheque fraud as a problem because they were not directly affected, except in the rare case that they were targeted for fraud.

2. *Mobilizing the players.* The process of change began with the commissioning of research study by the Home Office on the problem. This produced a well documented report (Levi et al., 1991) that: (a) made a very persuasive argument that substantial losses to cheque fraud were incurred by business; (b) identified points of vulnerability in the product and service design; and (c) outlined relatively straightforward steps that could be taken to reduce these losses. The Home Office followed up this report by sponsoring a series of meetings designed to foster cooperation among the major players in order to share information, develop databases of hot cheques and suspect bank accounts. The major players involved were banks, merchants, and law enforcement. The cooperative Association for Payment and Clearing Services (APACS) had already been set up, but now gained special impetus from the momentum that was built up from these series of meetings. The secondary businesses that resulted contributed to increased speed of verification of bankcards and increased vigilance on the part of POS staff. These secondary businesses monitor many billions of cheque

transactions daily. Finally, consumers were involved through consumer education initiatives conducted by all of the aforesaid players.

3. *Monitoring the effects of change*. All products and services remain vulnerable to criminal attack. However, their vulnerability is relative to other products and services and their points of vulnerability. In this case, it is likely that criminals turned away from cheque fraud to the easier and more lucrative credit card fraud.

Case Study 5: Smart Guns in the United States

It is hard to say when it first became apparent that there was something wrong with the design of handguns in the United States. The classic Colt .32, a 19th century best seller, was designed to be hard to shoot. In fact, it was nicknamed the "lemon squeezer." Arguably, this was an early indication that the designers of handguns recognized that a lethal weapon should not be made too easy to use. In 1880, alarmed by a child's death at the hands of a "lemon squeezer," Colt's Manufacturing Co. developed a gun with a grip safety lever that had to be pressed before pulling the trigger. They sold 400,000 of these guns (Fields, 2000). For another 100 years, the gun industry was seen as a venerable, truly iconic industry. But by 1999 (the year of the Columbine high school massacre in Colorado), gun manufacturers were under concerted attack to redesign their lethal products to make them safer to use.

The increasing pressure brought to bear on the gun industry can be traced through a series of defining events. The first was the attempted assassination of President Reagan and the serious wounding of his press secretary, James Brady, in 1981. This was followed by the formation of a powerful anti-gun lobby led by Handgun Control and the Brady Center to Prevent Handgun Violence. Their activities culminated in the Brady Bill and other legislation that banned assault weapons. This legislation galvanized the National Rifle Association (NRA) into action. Its funds helped deliver the Congress to the Republicans in 1994 and it successfully blocked every effort at gun control legislation thereafter for many years. Even in 1999, a year of horrendous mass shootings, the NRA was rated the most influential lobby group on Capitol Hill (Birnbaum, 1999).

In May 1998, another significant event occurred when an attorney from the Castano Group, lawyers who had won huge settlements from the tobacco companies on behalf of many U.S. states, contacted the Brady Center to Prevent Handgun Violence. Until this time, lawsuits brought against gun manufacturers claiming that their products had been "misused" (whether by children or criminals) had failed. Courts were reluctant to find a manufacturer liable for misuse of a product since this would create enormous problems for a large swath of industry. And after all, guns were *supposed* to be dangerous. However, once the connection was made with the successful litigation against the tobacco industry, the damage done by guns could be recast as a public health issue. The case was reinforced by statistics showing that most gun deaths in the United State were not criminally caused, but were accidents or suicides.

Consequently, in October 1998, the mayor of New Orleans, at the instigation of the Castano Group, filed a suit against 15 gun manufacturers and several local dealers. The manufacturers were characterized as displaying a "callous disregard" for the safety of children (Boyer, 1999). The city governments of Miami, St. Louis, and Chicago soon followed. Among the accusations made by the cities was that the manufacturers had deliberately suppressed research into "smart guns"—the technological solution to gun misuse.

These suits ultimately failed, but they served a valuable purpose in publicising the issues. The plaintiffs' lawyers understood that they had public opinion on their side: polls consistently have shown that the majority of Americans support registration of handguns, and that about half of those people who do not own handguns favour banning them altogether. The plaintiffs' lawyers busied themselves recruiting more cities and states to join their group of plaintiffs to sue the gun manufacturers. They knew from their experience with "Big Tobacco" that sooner or later a case would come up in which a jury would find in their favour.

In fact, the courts threw out a series of further cases. The first was in Chicago in 1998, and it concerned a police officer, shot on a public housing project, whose relatives charged that the gun manufacturers had nurtured a climate of violence by flooding the area with guns. The case was dismissed, but it was appealed. Next, Chicago Mayor Daley ordered a special police operation to uncover the distribution of guns and their flow into the illegal market in Chicago. This investigation was in preparation for a prosecution that would claim that the way in which gun manufacturers were flooding the market with guns was a "public nuisance"—an innovative legal argument and one with a stronger legal basis. However, the case was dismissed by a county court judge and, at about the same time, the New York Court of Appeals ruled in connection with a case brought on similar "public nuisance" grounds in Brooklyn that the gun industry cannot be held generally liable for shootings resulting from guns bought and sold illegally (Perez-Pena, 2001).

Although these cases were lost, the game is not over. In January 2002, the state appellate court of Illinois overturned the 1998 Chicago decision and ruled that gun makers could be sued for distributing guns in a way that makes it easy for criminals and juveniles to obtain them illegally (Butterfield, 2002). While the final decision on this issue is yet to come, or maybe will never be fully resolved, there is little doubt that the mounting number of cases is placing enormous pressure on the gun manufacturers. More than 30 cities and counties have sued, or are currently suing gun makers, and the plaintiffs' lawyers claim that the outcomes of these cases are relatively unimportant. Rather, it is the relentless process of litigation that will wear down the gun industry into submission and settlement and, in effect, the litigants have already won.

Ed Shultz, CEO of Smith and Wesson saw this coming, and, in a further significant event, broke ranks with the industry in 1998 when he accepted an invitation from President Clinton to attend a Rose Garden "photo-op" to publicize the administration's proposals for mandatory safety locks on handguns. (Shultz had caught the attention of the White House through a directive he had issued in 1997 that all Smith and Wesson handguns would be issued with a trigger-locking device.) This event marked the first split between the gun industry and the NRA, which had strongly opposed the Rose Garden meeting. Then in October 1999, Colt's Manufacturing Co. announced that it would no longer manufacture civilian handguns (Miller, 2000), claiming the costs of litigation were too great to justify continued production. This was ironic, since Colt's had expended considerable capital in developing and patenting a hi-tech smart gun that could be fired only by the person who owned it.

In December 1999, the Clinton administration said that it was preparing a class action suit on behalf of all 3,191 public housing authorities seeking to recover the costs of gun violence, and to force the design of safer firearms and restrict the flow of guns into illegal hands (Novak, 1999). In March 2000, Schultz, in an attempt to defuse class action suits by some 29 municipalities, made an agreement with the federal government that Smith and Wesson would include locks on all its handguns and research and implement "smart gun" technology.[44] This was a further break with the industry, for which Smith Wesson paid dearly. It was loudly vilified by customers and sales dropped dramatically, causing it to lay off 125 of its 725 employees (Seglin, 2001).

Finally, in an action that put pressure on Smith and Wesson's competitors, Andrew Cuomo, U.S. Secretary of Housing and Urban Development, and attorneys general from New York and Connecticut, announced that they would give preferential treatment in the procurement process to gun makers who adhered to the code of conduct that Smith and Wesson had signed. Seven gun manufacturers fought back with a lawsuit charging that Cuomo and his co-defendants were illegally trying to influence where law enforcement agencies bought weapons (Fields, 2000).

Events had not been kind to the NRA during this steady collapse of the gun industry. It had been taken over by a doctrinaire minority, which resulted in loud, uncompromising pronouncements. This was exacerbated by a huge mistake when it released a strident fund-raising letter that labelled officers of the Bureau of Alcohol, Tobacco and Firearms as "jackbooted government thugs." Unfortunately for the NRA, this letter was released the very day after the 1995 Oklahoma City bombing. Many politicians resigned their membership (former President George Bush among them). The NRA set about cleaning up its image, invited women to join, placed Charlton Heston (a famous movie star) at its head, championed gun safety education, and continued to lobby rural democrats and republicans. As a result, NRA membership has increased in recent years, though overall gun ownership has not. Male gun ownership has dropped well below 50%, and the manufacture and importation of guns has dropped some 20% since the 1970s (Birnbaum, 1999).

Case Study 6: The V-Chip for Televisions in the United States

The U.S. Congress first held hearings on the subject of television violence and its effects on children in 1952, but it was it was not until the mid-90s that any significant steps were taken to regulate program content. There had been much public debate during the 1994 election about the amount of violence on television, and considerable effort was made to evaluate the extensive research concerning the connection between television violence and the violent behaviour of children. While that connection remains controversial (ACLU, 1996), President Clinton in his 1996 State of the Union address challenged the television industry to develop a system for rating television programs. Less than one month later, Congress passed the bipartisan Telecommunications Act 1996, which among its many provisions contained two important ones relating to violence: (1) that industry submit a voluntary system of parental guidelines for rating television programming; and (2) that technology be installed in television sets to allow parents to block violent programs. The Federal Communications Commission (FCC) was charged with the responsibility for implementing these requirements, the second of which was eventually met by the invention of a device, the V-chip, that could read program ratings information from line 21 of the television transmission (the same line used for closed captioned information) and that also allowed a user-friendly way of blocking programs (Federal Communications Commission, 1997).

Advocates of the V-chip included individual politicians, as well as various parental and education groups including Kidsnet, the Kaiser Family Foundation, Children Now, the National Education Association, and the Academy of American Pediatrics. Opponents of the V-chip, or more accurately the ratings system, included the American Civil Liberties Union (ACLU), as well as a number of the most prominent broadcasting associations. The ACLU advanced the argument that the government was infringing on the First Amendment rights of broadcasters (ACLU, 1996) and the broadcast associations echoed this complaint (Stern, 1996). However, the essential resistance to the FCC was focused through Jack Valenti, the chairman of the Motion Picture Association of America (MPAA), on behalf of the entertainment media. He responded to the FCC demands for a rating system with an age-based ratings system that was basically the same as that already used by the motion picture industry (Fleming, 1996). Advocates criticized this solution on grounds that it was too vague. Eventually, Valenti (1997) returned with a revised six-category system that was a hybrid of age-based and content-based information, which the FCC accepted (Federal Communications Commission, 1998).

It is interesting to note that opposition to the FCC was focused almost entirely on the ratings system, not on the V-chip itself. Although some TV manufacturers claimed that the "best V-chip is a parent with a thumb to turn off the TV" (Kirkpatrick, 1997), their opposition soon disappeared when it came to estimating the cost of installing the device in TVs (about $3), and it became apparent they might even profit from this added device.

In May 1999, the FCC announced the formation of a special V-chip task force charged with working with consumer groups, industry, entertainment producers, and parents to ensure that the V-chip technology was properly implemented (Federal Communications Commission,

1999a). In June 1999, it was able to announce that all major television manufacturers had met or exceeded the deadline to have half of all new models with screens larger than 13 inches by July 1, 1999 equipped with the V-chip (Federal Communications Commission, 1999b). Six months later it announced that all the major broadcasting organizations were encoding their programming to work with the V-chip and that the broadcast networks had begun to run public service announcements to inform parents of the V-chip use and capabilities (Federal Communications Commission, 2000a). The FCC emphasized that these accomplishments had resulted from the task force working closely with industry and trade group partners.

The FCC was pleased to count consumers among its partners, but in fact consumers were only represented by particular interest groups. Direct input from the parents who would be expected to use the V-chip was not systematically sought. The only early indication of consumer attitude was reported in a survey conducted in 1997 (well after the V-chip had been set into motion), which found that, while 57% of the consumers favoured the idea of blocking objectionable programming, only 30% thought that the technology would do a lot of good for children (Gerson, 1997). Other indications soon appeared that consumers might not have been as impressed with the V-chip as were the politicians. In a 1998 poll, 65% of parents said they would block objectionable TV shows if they had a V-chip in their home, but 69% said they were not likely to buy a V-chip box to use with their current TV (Staff, 1999). One retail chain reported that its 18 stores had sold only two units between them, while others said they had sold none. In August 1999, the Kaiser Foundation reported that only 39% of consumers had heard anything at all about the V-chip (Rarey, 1999) and a year later reported that the situation had not improved. By this time, about 15 million V-chip equipped sets were either in homes or at retailers (Williams, 2000).

The FCC blamed the network broadcasters for failing to promote the V-chip, perhaps with some justification (Federal Communications Commission, 2000b). In the first three months of 2000, the four networks aired public service announcements promoting the V-chip a total of 59 times, roughly 90 seconds worth a week (Barnhart, 2000). However, in mid-2001, after they claimed to have made a special effort to promote use of the V-chip, the *New York Times* reported that the parental use of, or interest in the V-chip was still dismal, with only 7% of parents surveyed reporting that they ever used the V-chip to block programs (Rutenburg, 2001).

One reason for the V-chip's lack of success is that, because it is the same for all television sets, it breaks an important rule of retailing: always have a feature that one's competition does not have (Barnhart, 2000). Thus, there was no business reason to promote TVs by advertising their V-chip capability. It might be argued that if parents had demonstrated a strong interest in V-chip technology, then the industry would have developed competing ways to respond to such a demand. But when the manufacturers looked at sales data, and talked to their retailers, they understood that the V-chip was dead as a sales promotion device. It was clear that, even after substantial attempts to educate parents about the V-chip, they do not care about it and find little use for it.

Case Study 7: Redesign of Banknotes Worldwide

Governments monopolize the manufacture of currency in most modern nations,[45] though this was not always the case. For example, during the U.S. Civil War, there were some 1,600 state banks all printing their own notes, which created ideal conditions for counterfeiting. Detailed information on the losses due to counterfeiting is not freely available. It has been claimed by the U.S. Secret Service that the value of counterfeit U.S. currency *seized* has declined from $110 million in 1988 to $24 million in 1993. In addition, it estimated that the value of counterfeit money in *circulation* had risen from $11 million to $19 million (Houston, 1994). Considering the enormous number of banknotes in circulation, these amounts are tiny and are also low in comparison to the estimates of losses to other types of fraud (see credit card and check fraud case studies).

There are two reasons why there is relatively little counterfeiting: (1) governments devote special resources to tracking down and preventing counterfeit notes from getting into circulation,[46] and, (2) producing convincing copies of banknotes is particularly difficult and expensive because of the high level of protection the banknote printing industry has maintained concerning its trade secrets and manufacturing process.[47] This is remarkable when one considers that the organizations printing banknotes are a mixture of government and private companies that not only print for the domestic market, but also print notes for many other countries. Some of these companies also print other high security documents such as lottery tickets, and stock and bond certificates.

The first line of defence in security of the bank note is the design of clearly visible, *overt* features that ordinary people can easily detect. The rule of thumb here is that there should be as many as possible of these features at different levels, so that if one is defeated others will still prevail. They should also be durable and designed in such a way that there is room to add new features, should these become necessary. For example, the new Euro notes contain:[48]

1. *Security threads.* When the banknote is held up to the light, a dark line becomes visible.

2. *Watermarks.* When the note is held up to the light, picture and value of the banknote become visible.

3. *Foil holograms.* On the front of the low value notes there is a hologram foil stripe. When the banknote is tilted, the euro symbol and the value of the banknote appear.

4. *Iridescent strips.* These appear on the reverse side of the banknote. When the banknote is tilted under a bright light, the iridescent stripe shines and slightly changes colour

5. The higher denominations also have a *foil patch* and *colour shifting ink.*

In fact, Australia's use of polymer substrate banknotes moves even closer to the production of a self-authenticating bank note: security features that are so clear that anyone will immediately see them, yet they cannot be counterfeited (Curtis, 2000). The second line of defence is *covert* features, dependent on printing technology (deep engraving, latent images, fountain prints) and materials used (inks and paper or paper substitutes), that are not apparent to the ordinary user, or even the sophisticated counterfeiter, but which are readable by machines.

Many redesigns of currency have been forced by inflation,[49] so countries with weak economies are particularly hesitant to redesign their banknotes lest this be taken as a sign of impending devaluation. However, even countries with strong economies are reluctant to change their banknotes, because the value of their currencies depends heavily on public confidence, and this confidence depends on familiarity with the currency. This is why governments embark on extensive public education campaigns when they do introduce a new design (Reserve Bank of Australia, 2000) or a completely new currency such as the Euro (Staff, 1998b). The United States has a special problem because the dollar's strength makes it universally exchangeable and it is hoarded throughout the world.[50] When the U.S. government released its redesigned $100 bill in 1996, it had to take special steps to allay fears of the Russian government that the old notes would still retain their value (Federal Information Systems Corporation, 1996).

Governments redesign their banknotes to improve security for the following reasons:

1. Specific events act as "wake-up calls." For example, a so-called "supernote" surfaced in Lebanon in 1995, where it was estimated that (US) $2 billion in counterfeit notes had been manufactured (Skidmore, 1997). Similarly, a major counterfeit scare in Australia caused that country to embark on a complete revision of its banknotes (Australian Institute of Criminology, no date).

2. Losses due to counterfeiting have gradually risen to an unacceptable level through an increase in counterfeiting.

3. International pressure to update currencies that are perceived as too easily copied. This may have helped push the United States into changing its notes in 1995, the first change since 1929.[51]

4. A currency "arms race" among countries that are improving the security features of their banknotes. It is surely not a coincidence that the following countries all embarked on redesign of their banknotes (often the first changes in years) at around the same period, from the late 1980s through to the end of the 1990s: U.S.A.; Australia (Note Printing Australia, 2001; Australian Institute of Criminology, no date); Canada (Gazin, 2001); Japan (Staff, 2000a); Mexico (Gregory, 1994); United Kingdom;[52] Vietnam (Staff, 2000b); Switzerland (Studer-Walsh, 1995).

5. Changes in technology that make it easier for counterfeiters to copy banknotes. Recently, the most important technological innovations have been low-cost colour copiers and inkjet printers coupled with personal computers and scanners (Hecht, 1994). The advent of the colour copier, in particular, has had the important effect of directing the attention back to the importance of overt security features.

✦

Address for Correspondence: Ronald Clarke, School of Criminal Justice, Rutgers University, 123 Washington Street, Newark, NJ 07102 (e-mail: rvgclarke@aol.com).

Acknowledgments: We should like to thank Rob Guerette and Phyllis Schultze for their extensive help in obtaining the literature covered in this review. We are also grateful to Paul Ekblom, who managed the project on behalf of the Home Office. We benefited considerably from his unrivalled knowledge of criminogenic products.

NOTES

[1]See Ekblom (2000) for these and other examples of early product changes.

[2]See for example: www.designagainstcrime.org.uk; Erol et al., 2002

[3]Shover, 1996 shows how successive improvements made to safes in recent decades have led to the virtual extinction of safecracking.

[4]Theft and illegal use of these phones continues to be a problem in Britain (Harrington & Mayhew, 2002).

[5]BMW Australia was the world's first car manufacturer to spray all new cars (from September 2001) with up to 10,000 microdots each carrying the vehicle's unique VIN, though other manufacturers in Australia quickly followed BMW's lead (National Motor Vehicle and Theft Reduction Council, 2001b).

[6]The pain reliever OxyContin is a time-release tablet prescribed widely for terminally ill cancer patients and others suffering severe chronic pain. If bitten to break their seals, these pills can offer 10 times the strength of a single dose. The company making them, Purdue Pharma, has been asked to repackage them to prevent this misuse by drug addicts. The company has responded by limiting distribution of the tablets in large quantity packages, much as the sale of Aspirin in the U.K. is limited to no more than 30 tablets at a time. It is also developing a chemical additive (similar to that used previously with a time release drug) that will deactivate the narcotic once the seal of the capsule is broken (see http://www.injury-lawyer-network.com/oxycontin.htm).

[7]The New York City Taxi and Limousine Commission has for years "grappled with ways to make meters tamper proof" (Pierre-Pierre, 1995, p. B1).

[8]One example that recently came to light concerned changes contemplated for prepayment fuel meters and token vending machines after employees of the supplier were found to have tampered with them (Macharia, 1999).

[9]Hill, 1986; Staff, 1988. The need to replace coin meters was underlined at a crime prevention seminar held at 10 Downing Street in January 1986 (Tirbutt, 1986).

[10]In addition to incorporation of anti-counterfeiting measures, these documents need to be linked to an on-line database allowing real time access for enforcement agencies (Jill Dando Institute of Crime Science, 2002).

[11]See for example: www.scrip-safe.com

[12]Natarajan et al. (1996) describe a variety of modifications made to pay phones in drug-dealing areas, including blocks on incoming calls, blocks on coin operation at night and installation of phones with rotary dials (that cannot be used to call pagers).

[13]New car license plates have been adopted so that "it is virtually impossible to tell whether the driver comes from Bosnian-Serb Republic or from Muslim-Croat Federation" (BBC News Online, 1998). The old plates used the different scripts for Moslem, Serb or Croat regions, which led to danger for people travelling away from home.

[14]In some other cases, however, the link between the preventive action and the intended preventive benefit depends on untested assumptions. For example, monogramming a dressing gown might deter thefts by some hotel guests, but could increase thefts by others who wanted a souvenir. Again, despite assumptions to the contrary, the vast majority of parents might simply not bother to make use of the V-chip to prevent their children from watching violent TV programs.

[15]Bullet-proof partitions were made compulsory for all New York City licensed taxicabs in 1994. Various policing measures were also introduced at the same time. Robberies of cab drivers fell from 3,675 in 1993 to 1,089 in 1996 (Sullivan, 1997). This was a substantially greater drop than for robberies overall in the city during the same period, which fell from 86,001 to 49,670.

[16]See chapters in Felson and Clarke (1997).

[17]This is a more detailed classification than used by Lester (2001) in a review of product change for the Australian Institute of Criminology.

[18]See also: www.research.linst.ac.uk/dac

[19]As mentioned, these can be difficult to identify. Product tampering, for example, poses a threat to public safety, but it also presents a severe economic threat to retailers and manufacturers of foodstuffs and pharmaceuticals. In other cases, however, the main victims can be more easily determined. For example, the costs of car theft have been shown to fall mainly on the public, who must pay more for insurance and who can be severely inconvenienced by the loss of a vehicle (Field, 1993), while those of credit card fraud fall mainly on financial institutions, who generally indemnify the retailers and cardholders.

[20]The multiple jurisdictions in the United States sometimes result in legal confusion, as when the banks challenged the right of the New York City Council to impose regulations on ATM machines on the grounds that banks were subject to federal controls (Guerette & Clarke, 2003).

[21]Immobilisers are electronic devices that isolate two separate circuits that the vehicle requires to run. They come into automatic operation 20–40 seconds after the ignition is turned off. The EU is also working on a Standard to govern immobilisers that can be activated by remote control after a vehicle has been reported stolen.

[22]No formal evaluations of immobilisers have been published, though statistics from Western Australia (National Motor Vehicle Theft Reduction Council

[2001a]), the United States (Wollenberg, 2000) and the U.K. (Brown & Thomas, 2003; Vehicle Crime Reduction Action Team, 1999) suggest they are effective.

[23]A newspaper story carried the following headline: "Home Office study urges manufacturers to help curb a crime trend that leaves teenagers at particular risk" (Travis, 2002).

[24]These measures include: blocking of calls from phones reported stolen; blocking of calls associated with a stolen SIM card; pooling of information among the five companies about stolen phones and SIM cards (Baird, 2002).

[25]*Design against Crime* had four objectives (www.designagainstcrime.org): (1) constructing a database of best practice case studies; (2) developing teaching packages to support the teaching of design against crime from school through to degree level; (3) holding a design competition for higher education institutions; and (4) disseminating information and providing training on crime-resistant design for design professionals.

[26]An important exception is the report produced by IPPR with the support of business and reproduced in Chapter 2 of this volume. This clearly recognises that business products and services can cause crime and it argues that, consistent with Corporate Social Responsibility (CSR), businesses have an important role in crime prevention: "The case for a company to act is strongest the greater their product or service plays in causing crime, and the greater the contribution that the company could play in reducing the opportunity for crime, compared to the contribution of other players" (Hardie & Hobbs, 2002, p. 6).

[27]The U.S. Wireless Telephone Protection Act 1998 criminalized the use, possession, manufacture or sale of cloning hardware or software.

[28]A detailed discussion of this point, together with prescriptions for developing a knowledge base of crime prevention is offered by Ekblom (2002).

[29]Ignition interlocks are fitted to the cars of convicted drunk drivers in many jurisdictions in the United States with some degree of success (Beck et al., 1999; Marques et al., 1999a,b), but not into cars at manufacture.

[30]Economical devices have been patented that permit operation of spray cans only when tethered to an electrical outlet. This renders them useless for spraying graffiti, but still allows the spray cans to be used for legitimate painting jobs. Without a law banning the sale of ordinary spray cans, no market would exist for the devices (Ingram, 1996).

[31]www.Senter.nl/t&s/crimi

[32]The Home Office Police Scientific Development Branch allocated £5.5 million from March 2000 to fund a series of demonstration projects, "The Chipping of Goods Initiative," to show how property crime can be reduced through the retail supply chain using radio frequency identification (RFID) technology (Adams & Hartley, 2000).

[33]Pease (1998, p. 44) describes how new television sets could be protected by this technology: "Digital TVs have a uniquely identifiable microprocessor which can be interrogated remotely. Stolen TVs can thus be electronically deactivated, e.g., by using a Ceefax page containing the numbers of stolen TVs which is

scanned automatically at switch-on. If a TV finds its own number, it switches itself off, and remains unusable."

[34]Although Shepherd et al. (1990a) report that most glasses in pub assaults are shattered on impact with the victim, one of the writers has observed in 1950s Australia, the preference of breaking the glass or bottle first.

[35]Barasch v. The Bell Telephone Company of Pennsylvania, 201 E.D. 1990, 202 E.D.1990, S. Ct. Penn. (1992).

[36]Telecom Privacy Digest Fri, 03 May 1991, Volume 2, Issue: 051. Moderator: Dennis G. Rears. Today's topics: Twisting the phone system against the unwary; ACLU Position on Per-Line Blocking? Caller-ID—New Twist on an Old Capability; Blocking (http://www.infowar.com/iwftp/cpd/CPD-z-Telecom/V2_051.txt).

[37]These matters are covered in the following directives and recommendations:

European Parliament and the Council of the European Union Directive 95 // of the European Parliament and of the Council on the protection of individuals with regard to the processing of personal data and the free movement of such data (adopted by the European Council of Ministers 25 July 1995). This directive sets out a general framework for domestic private sector data protection legislation in the member countries.

Commission of the European Communities Amended proposal for a European Parliament and Council Directive concerning the protection of personal data and privacy in the context of digital telecommunications networks, in particular the Integrated Services Digital Network (ISDN) and digital mobile networks. This directive mandates minimum privacy protection standards for the provision of Caller-ID services.

Council of Europe Recommendation No. R (95) 4 on the protection of personal data in the area of telecommunications services, with particular reference to telephone services (adopted 7 February, 1995). This recommendation arose out of the Council of Europe Convention for the Protection of Individuals With Regard to Automatic Processing of Personal Data (No. 108 of 1981).

[38]Lambourne (1992). (It is estimated that 70% of all drugs sold in Africa are counterfeit.)

[39]http://www.watermarkingworld.org/intro.html

[40]Credit cards, bankcards, and cheque cards are a special case considered elsewhere in this report.

[41]Cheque and Credit Clearing Company is a subsidiary of APAC (Association for Payment Clearing Services): http://www.apacs.org.uk/

[42]Larabee (1999). (The annual cost of nationwide cheque fraud for the USA is estimated at $4 billion, compared to compared to $68 million for robbery.)

[43]http://www.printerm.com/fraud

[44]Technologies available for smart guns include simple trigger locks with keys, fingerprint-activated locks, a wristband or finger ring transponder activated lock, PIN requirements and smart displays that show when a gun is loaded and when it is locked.

[45]Though in Scotland banks still issue their own notes.

[46]In fact, most counterfeit notes do not reach circulation, but are seized before reaching the market (Federal Document Clearinghouse, 1994b). (Testimony July 13, 1994 by Robert J. Leuver).

[47]Federal Document Clearinghouse (1994a). (Testimony July 13, 1994 by Morris Weissman).

[48]http://www.bnb.be/EU/E/page6.htm

[49]Federal Document Clearinghouse (1994b). (Testimony July 13, 1994 by Robert J. Leuver).

[50]Federal Document Clearinghouse (1994b). (Testimony July 13, 1994 by Robert J. Leuver).

[51]Federal Document Clearinghouse (1994b). (Testimony July 13, 1994 by Robert J. Leuver)

[52]Aldersey-Williams (1998). (See also the Bank of England web site: http://bankofengland.co.uk).

REFERENCES

American Civil Liberties Union. (1996, February 29). *ACLU expresses concern on TV rating scheme; Says, voluntary system, is government backed censorship* (Briefing Paper #14). Retrieved from http://www.aclu.org/library/aavchip.html

Adams, C., & Hartley, R. (2000). *The chipping of goods initiative. Property crime reduction through the use of electronic tagging systems* (Police Scientific Development Branch, Publication No. 23/00). London: Home Office.

Aldersey-Williams, H. (1998). Money and the national soul. *New Statesman, 127*(4375), 41.

Anderson, R. (1998). On the security of digital tachographs. In J. J. Quisquater (Ed.), *Computer security, ESORICS 98: 5th European Symposium on Research in Computer Security, Louvain-la-Neuve, Belgium, September 16–18, 1998: Proceedings* (pp. 111–125). Berlin and New York: Springer. (Lecture Notes in Computer Science, #1485)

Australian Institute of Criminology. (n.d.). *Preventing the counterfeiting of Australian currency*. Retrieved from http://www.aic.gov.au

Australian Communications Authority. (1998, June 11). *Calling number display: Third report of the Austel Privacy Advisory Committee*. Retrieved from http://www.aca.gov.au/consumer/reports/cnd/intex.htm

Baird, R. (2002, February 9). Crusade to smash mobile phone muggers. *Daily Express*, pp. 1, 6.

Barasch v. The Bell Telephone Company of Pennsylvania. (1992). 201 E.D. 1990, 202 E.D.1990, S. Ct. Penn.

Barnhart, A. (2000, June 26). TV's educational efforts get so-so grades. *Kansas City Star.*

BBC News Online. (1998, February 2). *Bosnian licence for silence.*

Beck, K. H., Rauch, W. J., Baker, E. A., & Williams, A. F. (1999). Effects of ignition interlock license restrictions on drivers with multiple alcohol offenses: A randomized trial in Maryland. *American Journal of Public Health, 89*, 1696–1700.

Benham, M. (2002, January). Travel smart cards ready this year. *Evening Standard* (London), p. 2.

Bichler, G., & Clarke, R. V. (1996). Eliminating pay phone toll fraud at the Port Authority Bus Terminal in Manhattan. In R. V. Clarke (Ed.), *Preventing mass transit crime*. Crime Prevention Studies, Vol. 6. Monsey, NY: Criminal Justice Press.

Birnbaum, J. H. (1999, December 6). Under the gun. *Fortune*.

Boyer, P. J. (1999, May 17). Big guns. *New Yorker*, p. 64.

Braga, A., & Clarke, R. V. (1994). Improved radios and more stripped cars in Germany: A routine activity analysis. *Security Journal, 5*, 154–159.

Brantingham, P. L., & Brantingham, P. J. (1993). Environment, routine and situation: Toward a pattern theory of crime. In R. V. Clarke & M. Felson (Eds.), *Routine activity and rational choice*. Advances in Criminological Theory, Vol. 5. New Brunswick, NJ: Transaction Press.

Bridgeman, C. (1997). Preventing pay phone damage. In M. Felson & R. V. Clarke (Eds.), *Business and crime prevention*. Monsey, NY: Criminal Justice Press.

Brill, H. (1982). Auto theft and the role of big business. *Crime and Social Justice, 18*, 62–68.

Brown, R., & Thomas, N. (2003). Aging vehicles: Evidence of the effectiveness of new car security from the Home Office Car Theft Index. *Security Journal, 16*(3), 45–54.

Butterfield, F. (2002, January 3). Suit against gun makers gains ground. *New York Times*, p. A16.

Carroll, K. (1999, November 29). Caller ID takes to the Internet. *Telephony*.

Cerasale, J. (1999, November). Pertinent new and pending legislation for telephone marketers. *Call Center Solutions*.

Challinger, D. (1991). Less telephone vandalism. How did it happen? *Security Journal, 2*, 111–119.

Christie, R. (2000). *Targeting zero car theft: Lessons from the road safety field*. Paper presented at the conference of the Australian Institute of Criminology.

Clark, A. (2001, June 18). Security of the future. *Electronic Engineering Times*. Retrieved from www.electronicstimes.co.uk

Clarke, R. V. (1992). Deterring obscene phone callers: the New Jersey experience. In R. V. Clarke (Ed.), *Situational crime prevention: Successful case studies*. Albany, NY: Harrow and Heston.

Clarke, R. V. (Ed.). (1997). *Situational crime prevention: Successful case studies* (2nd ed.). Monsey, NY: Criminal Justice Press.

Clarke, R. V. (1999). *Hot products: Understanding, anticipating and reducing the demand for stolen goods*. Police Research Series, Paper 98. London: Home Office.

Clarke, R. V. (2000). Hot products: A new focus for crime prevention. In S. Ballintyne, K. Pease, & V. McClaren (Eds.), *Secure foundations. Key issues in crime prevention, crime reduction and community safety*. London: Institute for Public Policy Research.

Clarke, R. V. (2005). Seven misconceptions of situational crime prevention. In N. Tilley (Ed.), *Handbook of crime prevention and community safety*. Culompton, UK: Willan Publishing.

Clarke, R. V., Cody, R., & Natarajan, M. (1994). Subway slugs: tracking displacement on the London Underground. *British Journal of Criminology, 34*, 122–138.
Clarke, R. V., & Harris, P. M. (1992a). A rational choice perspective on the targets of autotheft. *Criminal Behaviour and Mental Health, 2*, 25–42.
Clarke, R. V., & Harris, P. M. (1992b). Auto theft and its prevention. In M. Tonry (Ed.), *Crime and justice: A review of research* (Vol. 16). Chicago, IL: University of Chicago Press.
Clarke, R. V., Kemper, R., & Wyckoff, L. (2001). Controlling cell phone fraud in the US: Lessons for the UK 'Foresight' prevention initiative. *Security Journal, 14*, 7–22.
Cohen, L. E., & Felson, M. (1979). Social change and crime rate trends: A routine activity approach. *American Sociological Review, 44*, 588–608.
Cole, S. (2001, February). Keep it real with security documents. *Business Forms, Labels & Systems*. Retrieved from http://www.bfls.com
Cornish, D., & Clarke, R. V. (Eds.). (1986). *The reasoning criminal*. New York: Springer-Verlag.
Cornish, D., & Clarke, R. V. (2003). In M. J. Smith & D. Cornish (Eds.), *Theory for practice in situational crime prevention*. Crime Prevention Studies, Vol. 16. Monsey, NY: Criminal Justice Press.
Crime Prevention Agency. (1997). *Car theft index 1997*. Communications Directorate. London: Home Office.
Crowe, T. D. (1991). *Crime prevention through environmental design: Applications of architectural design and space management concepts*. Boston: Butterworth-Heinemann.
Curtis, M. (2000, May 7–9). *The future of polymer notes*. Paper presented at the ICCOS Conference, San Diego, California.
Davis, R., & Pease, K. (2000). Crime, technology and the future. *Security Journal, 13*, 59–64.
Decker, J. F. (1972, August). Curbside deterrence: An analysis of the effect of a slug rejectory device, coin view window and warning labels on slug usage in New York City parking meters. *Criminology*, 127–142.
Department of Trade and Industry. (2000a). *Just around the corner: A consultation document*. Foresight Crime Prevention Panel. London: Department of Trade and Industry.
Department of Trade and Industry. (2000b). *Turning the corner*. Foresight Crime Prevention Panel. London: Department of Trade and Industry.
Design Council. (2002). *Design against crime*. London: Design Council.
DiLonardo, R. (1997, April). Radio frequency identification technology: everything old is new again. *Security Technology & Design*, 26–32.
Dodd, D. (1998). Making the cap fit. *Dairy Industries International, 63*(1), 21.
Ekblom, P. (1979). A crime-free car? *Research Bulletin*, 7, 28–30. London: Home Office.
Ekblom, P. (1986). *The prevention of shop theft: An approach through crime analysis*. Crime Prevention Unit, Paper 5. London: Home Office.
Ekblom, P. (1995). Less crime, by design. *Annals of the American Academy of Political and Social Science, 539*, 114–129.

Ekblom, P. (1997). Gearing up against crime: A dynamic framework to help designers keep up with the adaptive criminal in a changing world. *International Journal of Risk, Security and Crime Prevention, 2,* 249–265.

Ekblom, P. (1999). Can we make crime prevention adaptive by learning from other evolutionary struggles? *Studies on Crime and Crime Prevention, 6,* 27–51.

Ekblom, P. (2000). *Less crime, by design.* Lecture to Royal College of Arts.

Ekblom, P. (2002). From the source to the mainstream is uphill: The challenge of transferring knowledge of crime prevention through replication, innovation and anticipation. In N. Tilley (Ed.), *Analysis for crime prevention.* Crime Prevention Studies, Vol. 13. Monsey, NY: Criminal Justice Press.

Ekblom, P., & Tilley, N. (2000). Going equipped: Criminology, situational crime prevention and the resourceful offender. *British Journal of Criminology, 40,* 376–398.

Erol, R., Press, M., Cooper, R., & Thomas, M. (2002). Designing-out crime: Raising awareness of crime reduction in the design industry. *Security Journal, 15,* 49–61.

Evans, J. (2000, August 11). *Antitheft technology emerges.* Retrieved from Infoworld.com

Federal Communications Commission. (1997). *Commission proposes technical requirements to enable blocking of video programming based on program ratings.* News Report No. ET 97-8.

Federal Communications Commission. (1998). *Commission finds industry video programming rating system acceptable; Adopts technical requirements to enable blocking of video programming (the V-chip).* News Report No. GN 98-3.

Federal Communications Commission. (1999a, May 10). *FCC chairman William E. Kennard establishes task force to monitor and assist in the roll out of the V-chip to be chaired by commissioner Gloria Tristani.* News.

Federal Communications Commission. (1999b, June 9). *Commissioner Gloria Tristani commends manufacturers for meeting deadlines to install V-chips in televisions.* News.

Federal Communications Commission. (2000a, January 11). *FCC V-chip task force releases updated survey on the encoding of video programming.* News.

Federal Communications Commission. (2000b, April 4). *FCC commissioner Gloria Tristani urges TV networks to recommit to V-chip education efforts.* News.

Federal Document Clearinghouse. (1994a, July 13). *Testimony July 13, 1994 by Morris Weissman, chairman and chief executive officer U.S. Banknote Corporation, to the House Banking Committee on counterfeiting prevention.* FDCH Congressional Testimony. Capitol Hill Hearing Testimony.

Federal Document Clearinghouse. (1994b, July 13). *Testimony July 13, 1994 by Robert J. Leuver, executive director American Numismatic Association, to the House Banking Committee on counterfeiting prevention.* FDCH Congressional Testimony. Capitol Hill Hearing Testimony.

Federal Information Systems Corporation. (1996, March 27). *Press conference with Sergei Dubinin, chairman of the Central Bank and US Ambassador Thomas Pickering* *regarding introduction of the new US $100 bill.* Official Kremlin International News Broadcast. News/Current Events.

Felson, M. (1997). Technology, business, and crime. In M. Felson & R. V. Clarke (Eds.), *Business and crime prevention*. Monsey, NY: Criminal Justice Press.

Felson, M. (2002). *Crime and everyday life* (3rd ed.). Thousand Oaks, CA: Sage.

Felson, M., & Clarke, R. V. (Eds.). (1997). *Business and crime prevention*. Monsey, NY: Criminal Justice Press.

Felson, M., & Clarke, R. V. (1998). *Opportunity makes the thief.* Police Research Series, Paper 98. London: Home Office.

Field, S. (1993). Crime prevention and the costs of auto theft: an economic analysis. In R. V. Clarke (Ed.), *Crime prevention studies* (Vol. 1). Monsey, NY: Criminal Justice Press.

Fields, G. (2000, April 27). Gunmakers fight back. *USA Today*, p. 1.

Fleming, H. (1996, December). Valenti delivers the V-chip code. *Broadcasting and Cable.*

Foley, B. (1993). Taking the tamperers to task. *Soap, Perfumery & Cosmetics, 66*(4), 45.

Food and Drug Administration. (1992). *Requirements of laws and regulations enforced by the U.S. Food and Drug Administration.* Retrieved from www.fda.gov

Food and Drug Administration. (1999). *Guidance for industry: Container closure systems for packaging human drugs and biologics.* U.S. Department of Health and Human Services. Retrieved from www.fda.gov/cder/guidance/index.htm

Forbes, G. (2000). *Immobilising the fleet.* Paper presented at the conference of the Australian Institute of Criminology conference.

Garland, D. (1996). The limits of the sovereign state: strategies of crime control in contemporary society. *British Journal of Criminology, 36*, 445–471.

Gazin, G. (2001, January 31). New Canadian banknotes to stop counterfeiters new technology provides numerous anti-copy features. *The Edmonton Sun*, p. 55. See also the Bank of Canada web site: http://www.bankofcanada.ca

Gerson, R. (1997, October 27). V-chip: Consumers speak. *Twice, 12*(24), 6.

Gialamas, J. (1997, Winter). 100% pure art. *Forbes*, New York.

Goldsec. (1999). *Anatomy of a secure cheque: Microform 2000 security cheques.* Retrieved from http://www.goldsec.com/AnatomyofaSecureCheque.html

Gould, L. C. (1969). The changing structure of property crime in an affluent society. *Social Forces, 48*, 50–59.

Graham, K. (1984). Determinants of heavy drinking and drinking problems: The contribution of the environment. In E. Single & T. Storm (Eds.), *Public drinking and public policy.* Toronto: Addiction Research Foundation.

Graham, K., & Homel, R. (1997). Creating safer bars. In M. Plant, E. Single, & T. Stockwell (Eds.), *Alcohol: Minimising the harm: What works?* London: Free Association Books.

Greenman, C. (2001, July 5). Connecticut takes action against rental-car tracker. *New York Times*, p. G3.

Gregory, S. (1994, October 30). Under monetary overhaul, Mexico currency has new look. *San Diego Union-Tribune*, p. A18.

Guerette, R., & Clarke, R. V. (2003). Product life cycles and crime: Automated teller machines and robbery. *Security Journal, 16*, 7–18.

Hall, C. (1990, August 1). Caller-ID technology stirs up concern. *The San Francisco Chronicle*, p. C5.

Hardie, J., & Hobbs, B. (2002). *Partners against crime: The role of the corporate sector in tackling crime*. London: Institute for Public Policy Research.

Harrington, V., & Mayhew, P. (2002). *Mobile phone theft*. Home Office Research Study, No. 235. London: Home Office.

Harris, P. M., & Clarke, R. V. (1991). Car chopping, parts marking and the Motor Vehicle Theft Law Enforcement Act of 1984. *Sociology and Social Research, 75*, 228–238.

Hazelbaker, K. (1997). Insurance industry analyses and the prevention of motor vehicle theft. In M. Felson & R. V. Clarke (Eds.), *Business and crime prevention*. Monsey, NY: Criminal Justice Press.

Hecht, J. (1994). U.S. counts cost of computer counterfeiting. *New Scientist, 141*(1907), 19.

Hesseling, R. B. P. (1994). Displacement: A review of the empirical literature. In R. V. Clarke (Ed.), *Crime prevention studies* (Vol. 3). Monsey, NY: Criminal Justice Press.

Highway Loss Data Institute. (1996). *Insurance theft report: 1993–95 passenger vehicles*. Arlington, VA: Highway Loss Data Institute.

Hill, N. (1986). *Prepayment coin meters: A target for burglary*. Crime Prevention Unit, Paper 6. London: Home Office.

Hilts, P. (1982, November 5). FDA issues rules requiring tamper-resistant drug packaging. *The Washington Post*, p. A2.

Hough, M., & Tilley, N. (1998). *Getting the grease to the squeak: Research lessons for crime prevention*. Crime Detection and Prevention Series, Paper 85. Police Research Group. London: Home Office.

Houghton, G. (1992). *Car theft in England and Wales: The Home Office car theft index*. Crime Prevention Unit Series, Paper 33. Police Research Group. London: Home Office.

Houston, P. (1994, May 11). Money; Big chunk of change in design of greenbacks appears likely. *Los Angeles Times*, Section A5, 4.

Hyman, P. (1983). Legal considerations of tamper-resistant packaging for cosmetics. *Soap–Cosmetics–Chemical Specialties, 59*, 56.

Ingram, C. (1996, April 4). Graffiti bill would make spray cans less portable. *Los Angeles Times*, pp. A3, A14.

Jill Dando Institute of Crime Science. (2002, February 22). *Crime prevention and the UK vehicle registration and licensing system*. University College, London: Jill Dando Institute for Crime Science.

Karmen, A. A. (1981). Auto theft and corporate irresponsibility. *Contemporary Crises, 5*, 63–81.

Kasindorf, M., & Fields, G. (2000, April 27). Reluctant industry pursues smart guns. *USA Today*, pp. 1, 10a.

Keck, B. (1993). US Government lays down the law. *Packaging Week*, *8*(37), 27.

Kennedy, D. B., & Hupp, R. T. (1998). Apartment security and litigation: Key issues. *Security Journal*, *11*, 21–28.

Kirkpatrick, J. (1997, July 23). V-chips may be 2 years away. *Pittsburgh Post Gazette*.

Lake, M. (2001, August 2). Tweaking technology to stay ahead of the film pirates. *New York Times*, p. G9.

Lambourne, J. (1992). Tampering: Industry fights back. *Packaging Week*, *8*(23), 30.

Larabee, M. (1999, May 13). Banks battle high costs of cheque fraud. *The Oregonian*.

LaVigne, N. (1994). Rational choice and inmate disputes over phone use on Rikers Island. In R. V. Clarke (Ed.), *Crime prevention studies* (Vol. 3). Monsey, NY: Criminal Justice Press.

Lester, A. (2001). *Crime reduction through product design*. Trends and Issues No. 26. Canberra, Australia: Australian Institute of Criminology.

Levi, M., Bissell, P., & Richardson, T. (1991). *The prevention of cheque and credit card fraud*. Crime Prevention Unit, Paper 26. London: Home Office.

Levi, M., & Handley, J. (1998). *The prevention of plastic and cheque fraud revisited*. Home Office Research Study No 182. London: Home Office.

LiCalzi O'Connell, P. (2001, May 14). For consumer goods producers, its not so bad to be behind bars. *New York Times*.

Macharia, J. (1999, February 9). Tanzania firm probes meter fraud. *Business Day* (South Africa), p. 14.

Mansfield, R., Gould, L. C., & Namenwirth, J. Z. (1974). A socioeconomic model for the prediction of societal rates of property theft. *Social Forces*, *52*, 462–472.

Markus, C. L. (1984). British Telecom experience in payphone management. In C. Levy-Leboyer (Ed.), *Vandalism: Behaviour and motivations*. Amsterdam: Elsevier.

Marques, P. R., Voas, R. B., Tippetts, A. S., & Bierness, D. J. (1999a). The Alberta Interlock Program: The evaluation of a province-wide program on DUI recidivism. *Addiction*, *94*, 1849–1859.

Marques, P. R., Voas, R. B., Tippetts, A. S., & Bierness, D. J. (1999b). Behavioral monitoring of DUI offenders with the Alcohol Ignition Interlock Recorder. *Addiction*, *94*, 1861–1870.

Mason, T., & Little, A. (1997, April 12). *Labour pledges three-pronged crackdown on violent drinkers*. Press Association Newsfile.

Miller, J. J. (2000, April 17). Shot dead. How Colt was done in. *National Review*.

Monaghan, P. (1997, November 21). The scientist behind the V-chip. *The Chronicle of Higher Education*, *44*(13), A9.

Mueller, G. O. W. (with J. Shames & J. Chazen). (1971). *Delinquency and puberty: Examination of a juvenile delinquency fad*. Criminal Law Education and Research Center, Monograph Series, Vol. 5. New York: New York University.

Murphy, M. (1991). What price a reputation? *Packaging Week*, *7*(26), 6.

Natarajan, M., Clarke, R. V., & Belanger, M. (1996). Drug dealing and pay phones: The scope for intervention. *Security Journal*, *7*, 245–251.

National Motor Vehicle Theft Reduction Council. (2001a). *Immobilisers and Western Australia's declining theft rate.* Theft WATCH, Vol 1, June. Melbourne, Australia: National Motor Vehicle Theft Reduction Council. (See also: www.carsafe.com.au)

National Motor Vehicle Theft Reduction Council. (2001b). *BMW leads the fight against professional thieves.* Theft TORQUE, Special edition, July. Melbourne, Australia: National Motor Vehicle Theft Reduction Council. (See also: www.carsafe.com.au)

Neeley, D. (1998). Capping cart theft. *Security Management, 42,* 21–22.

Notarangelo, R. (1997, November 15). Tougher glasses to beat pub yobs. *Daily Record,* p. 2.

Note Printing Australia. (2001). *The world standard in security.* Retrieved from www.noteprinting.com/sc03world/sc031/body.html

Novak, V. (1999, December 20). Enter the big guns. *Time,* p. 59.

NRMA. (1987). Car theft: Putting on the brakes. Proceedings of Seminar on Car Theft, May 21, 1987. Sydney, Australia: National Roads and Motorists Association and the Australian Institute of Criminology.

Padula, D. (2000). *Secure identification of motor vehicles.* Paper presented at the conference of the Australian Institute of Criminology.

Paradise, P. (1995). Signal theft. *Electronics Now, 66,* 35 (7 pages).

Pease, K. (1997). Predicting the future: The roles of routine activity and rational choice theory. In G. Newman, R. V. Clarke, & S. G. Shoham (Eds.), *Rational choice and situational crime prevention: Theoretical foundations.* Dartmouth, UK: Ashgate.

Pease, K. (1998). Changing the context of crime prevention. In C. Nuttall (Ed.), *Reducing offending: An assessment of research evidence on ways of dealing with offending behaviour.* Home Office Research Study, No. 187. London: Home Office.

Pease, K. (2001). *Cracking crime through design.* London: Design Council.

Perez-Pena, R. (2001, April 27). State court sides with gun makers in liability case. *New York Times.*

Pierre-Pierre, G. (1995, July 19). Twist a dial and cab fare soars, police say. *New York Times,* p. B1.

Plastiras, J. (1998, January 19). Changeover to electronic postage meters offers more security, but at cost to firms. *Capital District Business Review.*

Poyner, B., & Warne, C. (1988). *Preventing violence to staff.* London: Health and Safety Executive, Department of Employment.

Privacy Rights Clearinghouse. (2000). *Fact Sheet # 19: Caller-ID and my privacy: What do I need to know?* Utility Consumers' Action Network. Retrieved from http://www.privacyrights.org

Rarey, M. (1999, August 2). V-chip investment. *Insight on the News, 15*(28), 41.

Rawls, R. L. (2001). Contraband diamonds: World community seeks technology to identify diamonds from regions in conflict. *Chemical and Engineering News, 79,* 35–36.

Reserve Bank of Australia. (2000, July 20). *Introducing polymer banknotes—public communication.* Information Note.

Rhodes, W., Norman, J., & Kling, R. (1997). *An evaluation of the effectiveness of automobile parts marking on preventing theft.* Unpublished report prepared for the National Institute of Justice by Abt Associates, Inc.

Robinson, B. (1986). New tampering tragedy: Where do we go from here? *Drug Topics, 130,* 90.

Rutenburg, J. (2001, July 25). Survey shows few parents use TV v-chip to limit children's viewing. *New York Times,* p. 1.

Seglin, J. L. (2001, March 13). When good ethics aren't good business. *New York Times,* Business Section, p. 4.

Shepherd, J. P., Price, M., & Shenfine, P. (1990a). Glass abuse and urban licensed premises. *Journal of Royal Society of Medicine,* 176–177.

Shepherd, J. P., Shapland, M., Irish, M., Scully, M., & Leslie, I. J. (1990b). Pattern, severity and aetiology of injury in assault. *Journal of Royal Society of Medicine, 83,* 75–78.

Sherman, L. W., Gottfredson, D., MacKenzie, D., Eck, J., Reuter, P., & Bushway, S. (Eds.). (1997). *Preventing crime: What works, what doesn't, and what's promising.* Office of Justice Programs Research Report. Washington, DC: U.S. Department of Justice.

Shover, N. (1996). *Great pretenders: Pursuits and careers of persistent thieves.* London: Westview Press/Harper Collins.

Skidmore, D. (1997, March 25). After a year, new $100 bills a success; security features—cutting. *Austin American-Statesman,* Business Section, p. D8.

Slawson, M. (1997). *Caller-ID basics.* Intertek Testing Services. Retrieved from www.testmark.com/callerid

Smith, M. J., Clarke, R. V., & Pease, K. (2002). Anticipatory benefits in crime prevention. In N. Tilley (Ed.), *Analysis for crime prevention.* Crime Prevention Studies, Vol. 13. Monsey, NY: Criminal Justice Press.

Southall, D., & Ekblom, P. J. (1985). *Designing for car security: Towards a crime-free car.* Crime Prevention Unit Paper 4. London: Home Office.

Staff. (1982, October 25). Tylenol's boost to safer packaging. *Business Week,* p. 37.

Staff. (1983). Food, packaging industries react fast in wake of Tylenol tampering. *Quick Frozen Foods, 45,* 42.

Staff. (1988, April 6). UK locks out the cash raiders. *Financial Times,* p. 3.

Staff. (1989). Tamper-evidence: Consumers have come to expect tamper-evidence for drugs, foods. *Packaging, 34*(5), 16.

Staff. (1990, April 7). GM weighs changes to foil thieves. *St. Louis Post-Dispatch,* p. 11a.

Staff. (1994, November 29). Deluxe to combat cheque fraud. *PR Newswire.*

Staff. (1997, November 15). Stronger glasses urged to reduce injuries in pubs. *The Herald* (Glasgow), p. 7.

Staff. (1998a, February 20). Home Office: Government crackdown on drunken yobs. *M2 Presswire.*

Staff. (1998b, July 22). EU to draw up strategy to combat counterfeit euro banknotes, coins. *Extel Examiner.*

Staff. (1999, February 4). Parents slow to buy the V-chip. *Chicago Tribune.*

Staff. (2000a, July 17). 2,000 yen note keeps printers busy. *The Nikkei Weekly,* Economy & Politics, p. 6.

Staff. (2000b, September). Vietnam issues new banknotes. *The Saigon Times Magazine.*

Stanford Research Institute. (1970). *Reduction of robbery and assault of bus drivers.* Technological and Operational Methods, Vol. 3. Stanford, CA: Stanford Research Institute.

Stephens, G. (1998). The rules of cheque fraud liability are changing: Are you prepared? *Business Credit, 100*(5), 23.

Stern, C. (1996, February 12). Broadcasters plotting V-chip legal strategy. *Broadcasting and Cable.*

Stillwell, E. (1989). Strategies for foiling tamperers. *Packaging, 34*(7), 39.

Stoffer, H. (1997, August 18). Big 3 says parts marking law is too costly, fails to cut theft. *Automotive News,* p. 4.

Studer-Walsh, M. (1995). A new look for the Swiss franc. *Swiss Business, 6,* 62.

Sullivan, B. (2001, December 27). *Retail gift cards often unprotected.* Retrieved from www.msnbbc.com/news/598102.asp?cp1

Sullivan, J. (1997, January 7). Judge limits police stops of taxicabs. *New York Times,* p. 25.

Svensson, B. (1982). A crime-prevention car. In E. Kulhorn & B. Svensson (Eds.), *Crime prevention, research and development division* (Report No. 9). Stockolm: The National Swedish Council for Crime Prevention.

Telecom Privacy Digest. (1991, May 3). *Moderator: Dennis G. Rears Today's topics: Twisting the phone system against the unwary; ACLU position on per-line blocking? Caller-ID—New twist on an old capability; Blocking.* Volume 2, Issue 051. Retrieved from http://www.infowar.com/iwftp/cpd/CPD-z-Telecom/V2_051.txt

Teresko, J. (1983, February 21). Tapping profits from tamper-proof rules. *Industry Week,* p. 72.

Tirbutt, S. (1986, June 23). Fuel tokens plan would thwart meter thefts. *The Guardian.*

Travis, A. (2002, January 8). Home Office study urges manufacturers to help curb a crime trend that leaves teenagers at particular risk. *The Guardian.*

Valenti, J. (1997). *Letter to FCC from MPAA, NAB and NCTA.* Also endorsed by AMA, ACP, APA National Association of Elementary School Principals, NEA and National PTA. Retrieved from http://www.fcc.gov/vchip/revprop.html

Van Dijk, J. (1994). Understanding crime rates: On the interactions between rational choices of victims and offenders. *British Journal of Criminology, 34,* 105–121.

Vehicle Crime Reduction Action Team. (1999). *Tackling vehicle crime: A five-year strategy.* Communication Directorate. London: Home Office.

Webb, B. (1997). Steering column locks and motor vehicle theft: Evaluations from three countries. In R. V. Clarke (Ed.), *Situational crime prevention: Successful case studies* (2nd ed.). Guilderland, NY: Harrow and Heston.

Webb, B. (1994). Preventing plastic card fraud in the UK. *Security Journal*, 7, 23–25.

WHICH? (1991, February). Cars at risk. *WHICH? Report on Car Security*, pp. 107–109.

Wilkes, P., & Costello, A. (2000). *The road to nowhere: The evidence for travelling criminals*. Home Office Research Study No. 207. London: Home Office.

Williams, K. (1998, December). The number's up. *Policing Today*, pp. 18–19.

Williams, S. (2000, August 15). V-chip? What's a V-chip? Few know the answer. *Star Tribune*, p. 1E.

Wise, J. (1982, September). A gentle deterrent to vandalism. *Psychology Today*, *16*, 31–38.

Wollenberg, Y. C. (2000). High-tech devices cut down on car snatching. *Medical Economics*, 77, 21.

Partners Against Crime: The Role of the Corporate Sector in Tackling Crime

by

Jeremy Hardie

and

Ben Hobbs
Institute for Public Policy Research

Abstract: *What companies produce, and the services they provide, often create significant opportunities for crime. Companies are better placed than anyone else to reduce those opportunities and that is the basis of their responsibility to do so. Their contribution to crime prevention is of enormous potential. This chapter sets out to answer three questions. First, what are the limits of a company's responsibility to prevent the criminal misuse of its products and services? Second, where does its responsibility end and that of the state begin? Third, what public policy framework—law, regulations, incentives, and guidance—should government set in place to encourage and persuade companies to mainstream crime prevention thinking at the design stage of their products and services and to redesign them if the crime potential emerges after the product or service is already on sale? It is argued that companies' responsibility for tackling crime can, but does not always, fit with the business case (that is, it may not always be in the direct financial interests of the company to contribute). But companies, and their staff in particular, are motivated by* *more than profit. The chapter concludes that companies' contribution to crime prevention is very significant, but that there has been surprisingly little recognition by government or the public of the contribution that they have made.*

1. INTRODUCTION

This paper is about what we have to do to tap the full potential of companies' contribution to crime reduction. It argues that their contribution is of central importance to the fight against crime, and that public policy has neglected it.

It is not only that it would be good to have more help from business to supplement what the state does. And it is not about companies just providing money. That sort of help, like corporate sponsorship of literacy program, is certainly valuable. But often what companies can contribute is marginal. They are peripheral both to the creation of such problems as illiteracy and to its reduction. In a straightforward sense, it is not really their problem.

It is quite different in the case of crime. Businesses can and should be central to its reduction. What companies produce and provide can generate the opportunity for crime, and they are much better placed than anyone else to reduce those opportunities. Indeed, it may only be the company that can do so. That is the basis of their responsibility.

What companies can do to prevent crime should not be seen as just a *pro bono* activity like sponsoring an orchestra. As we shall see, crime prevention can and should be integral to the design, marketing and security of the goods and services that they provide. It is pro bono in the sense that it does good. But it is more than that. It is a responsibility that only they can fulfill.

That is the claim that this paper sets out to make good. Our focus is not crimes against business itself, like shoplifting, or crimes that are marginal to public concern. Companies can have a significant impact on the crimes which impinge most on the daily life of citizens and which the police can find most difficult to address: street crime, which can often involve violence; stolen handbags in restaurants; theft of TVs and videos at home. We mean unsafe streets and car parks, facilities provided by companies that can prove too hospitable to those intent on crime. And we mean the offences that occur because a service provided by a company is abused, like the fights which break out when too much alcohol has been consumed in pubs and clubs; or illegal entry to the country via Eurostar (the company that operates trains through the Channel Tunnel between England and France).

These are among the crimes where companies can make all the difference. We know that because of hard facts about what companies achieved in the 1990s, when car theft fell sharply because car manufacturers introduced immobilizers, when credit card fraud was tackled effectively by the card providers, and shopping centers were made safer through installation of closed-circuit television (CCTV). These are real cases that are set out below. This is not merely a theoretical analysis.

The paper is not an attempt to scapegoat companies, nor to dump on them responsibility for solving crime. But it is central to our argument that we—not just companies but citizens generally—cannot expect the criminal justice system to deal with crime on its own. Crime can only be reduced if everyone—the state, the public, companies, local authorities, individuals and voluntary organizations—recognize what they have to do and do it.

This does not mean that companies have to take responsibility for every aspect of crime control. This paper is, in part, about the division of labor between the state and companies, and there are many areas of crime prevention where companies cannot and should not be expected to take responsibility. Nor does it mean forgetting that the prime responsibility for crime lies with the offender. But it does mean that every business has to exercise some responsibility for preventing crime, from the market stallholder to the multinational, because without them crime cannot be controlled.

It follows, however, that the state must play its part as well, and that means dealing with crimes against business: from shoplifting to fraud and arson. Crimes against business can be a low priority for government and the police. The Business Crime Survey published by the British Chambers of Commerce in December 2001 found 49% of business crimes were not reported to the police because of a lack of confidence in the police response. Yet it estimated the cost of crimes against business at £18.8 billion a year, just over half of businesses having been the victim of crime over the past year.[1] It is a central argument of this paper that crimes against business should not be neglected because they are seen as "victimless" or because companies are thought rich enough not to deserve protection.

Citizens and households, as a result of their day-to-day experiences, have developed improved crime consciousness: a way of thinking and acting which takes into account the daily impact of crime (for example, locking your house or the car door). This personal responsibility goes beyond co-operating with the police by providing access to evidence after the event.

So too, companies have taken some responsibility for reducing the crimes of which they are themselves victims, by training staff to reduce the opportunities for crime and to detect offenders, by introducing security guards, strengthening physical protection such as locks and gates, and CCTV surveillance.

It is to be expected that companies will indeed take seriously the crimes of which they are the victim and do what they can to prevent them. Their motivation is clear. But this paper is not primarily about crimes against business. Nor is it about the purely *pro bono* involvement of companies in the criminal justice system, such as the encouragement of employees to become magistrates or prison visitors. Our focus is the analytically and practically more difficult middle ground: where the product or service produced provides an opportunity for crime, but the crime costs the company little or nothing. Here, the motivation of the company to act is less clear.

Limits of Responsibility

Clandestine entry to the U.K. would be much reduced if no truck ever carried a stowaway across the channel. But is it right that the state should hold transport companies responsible for unknowingly assisting in this offence, and punish transport companies if they do so? If we do accept that companies bear some responsibility for crime prevention, where does their responsibility end and that of the state begin?

What about alcohol? Much hooliganism and violence is associated with drink. These crimes would be reduced if the drink companies, the pubs and the clubs, did not produce and market alcoholic drinks, and make them available in accessible and attractive places late at night. Does that make the drink companies responsible for hooliganism? Should they be required to do something about it? What? How far should they be required to go?

A third example: Fewer cars get stolen if immobilizers are fitted. So now all new cars have to be so equipped. How did that happen? Why did the car manufacturers do it? Did it pay them to do so or were their motivations more complex? Why did legislation come last, when the problem was already on its way to solution? These are among the questions that we attempt to tackle.

Role of the Law

The example of the cross-channel truckers demonstrates that, once Government has decided that companies in a particular sector should act, a central question is the appropriate role of the law in requiring them to take action to address the crime potential in their products and services. A simplistic answer would be that we should identify the ways in which companies can make a contribution, and make them responsible for so doing by imposing a legal requirement on them.

One of the purposes of our study is to see what role the law, as compared with voluntarism, can and should play in successful crime reduction. The law can be used in at least two different ways. It could impose direct obligations on particular companies, as on airlines and Eurotunnel truckers in relation to clandestine entrants: it could require transparency in company reporting about the crime prevention steps that they have taken. Certainly, as we shall show in the case of car immobilizers, the law has a role to play in this way. Or it could play an enabling role—creating local crime reduction partnerships which companies are able to join, for instance. The Crime and Disorder Act 1998 led to many good local Crime and Disorder Reduction Partnerships, which include businesses, because of the statutory duty it imposed on local authorities in the U.K. to formulate crime strategies in partnership with others. In other cases, however, the law has played a negligible role.

Designing out Crime

A much-discussed example of company involvement in crime prevention is the notion of designing-out crime—thinking about new and different ways of designing products and services to reduce the opportunity available to the criminal to commit an offence. This includes anticipating threats—what has been called "thinking thief" (Ekblom, 2000)—and shaping design early enough to cope with them. Secure design concepts have long been used to enhance levels of protection, particularly to deal with crime against business. These have ranged from glass screens in banks to the electronic tagging of clothes and door entry codes. But our interest is to explore how manufacturers and retailers can be encouraged to think about the consequences of their activities for those crimes which do not directly affect them—to design products and services that are not only convenient and attractive, but also secure to use.

A modest but striking example is hooks for handbags on tables in cafes, which makes it harder for bag snatchers. Or the shatterproof glasses that make it less likely that Saturday night drinking will degenerate into grievous bodily harm. Retailers of personal computers have fitted their computers with filters to screen out pornography to protect child users. There is no problem in identifying improvements that are technically possible. The hard question is how manufacturers, cafe proprietors, publicans and retailers can be convinced that they ought to invest money to do so. Why should they?

Learning from Success

The case studies described in the second part of this chapter give details of three major successes in this middle ground. They show that companies have been at the heart of reducing car theft, credit card crime, drunken disorder, and shopping centre disturbances. They show the strengths and weaknesses of these initiatives, and limitations of the public policy framework in which they took place. It is important to start with such successes to establish the major contribution that companies can bring, and then to focus on how we can make more of their contribution.

It is important to note that these successes have not, certainly until recently, been the result of clear public policy initiatives. There has been much reliance on ad hoc co-operation between companies in a particular industry, and between companies as a group and central government, the criminal justice agencies and local authorities. Surprisingly, neither the Home Office (2001) White Paper—*Criminal Justice: The Way Ahead*—nor the Department of Trade and Industry's (DTI) *Business and Society* (2001) makes any significant reference to the role of business in reducing crime. Nor does the 1998 Crime and Disorder Act, despite its emphasis on local Crime and Disorder Reduction Partnerships (CDRPs). It is the substance of that neglect which this paper sets out to put right.

This paper thus reflects a number of themes.

First, that we all, companies included, have a shared objective of reducing crime. Thirty-six thousand crimes are committed each day in the U.K., creating fear, insecurity, and suffering and an estimated social and economic cost of £60 billion each year (Home Office, 2001b).

Second, there is a need for citizens, including corporate citizens, to take responsibility for doing what they can and should do about the problem of crime, and not passively look to the state to solve all their problems.

Third, it is also about partnership, that each of the players has a role which complements and must be coordinated with the others in a system of mutual co-operation. The lead partner will vary from case to case. There is no presumption that it will be the state, or any particular agency such as the police, which will be the most influential member.

Fourth, the ethos of partnership can be undermined by the legal enforcement of obligations and the authoritarian language that can accompany it. Accordingly, our public policy conclusions are heavily weighted on the side of voluntarism or soft regulation as against coercive legislation. This does not rule out enabling legislation. Nor does it preclude coercive legislation once voluntary co-operation has established what constitutes best practice, and there is a consensus that there should be rules to ensure that all companies conform. But the state should be wary of using coercive legislation, because it undermines the ethos of partnership.

Fifth, our conclusions are based on evidence. We provide three detailed case studies showing what companies already contribute to crime reduction. Their purpose is to put it beyond doubt that companies are central to crime prevention, because there are some steps that only companies can take and what they can do is very effective. And they point to clear lessons for the future development of public policy.

Finally, we emphasize that our policy proposals amount to generalizing the techniques and measures that in an ad hoc way have been around successfully for years.

We start in the next section by discussing the question of corporate social responsibility, and the reasons why companies are, and could become, motivated to get involved in crime prevention. In section 3 we set out the current public policy framework and recent operational initiatives, before turning in section 4 to the case studies. Section 5 examines the lessons to be drawn from our case studies and the further examples we have cited, and from the research evidence in this area. Finally, we set out recommendations for future action.

Like others, we see the development of environmental awareness within companies, and acceptance of their shared responsibility for the environment, as an illuminating precedent for how successfully attitudes can change. It takes time, and although it may include legislation as part of the armory of change, that is only one part, and works best at the end when popular, customer, and corporate opinion have already done most of the work. The result in the case of the environment is clear and gratifying. Everyone in Britain, including the private sector, now believes that we should take the future of the environment seriously, and that companies must take some responsibility for protecting it. The details remain a matter of dispute but the general principle is no longer contentious. One aim in this paper is to bring about a similar change in business attitudes towards responsibility for crime prevention.

The role of companies in tackling crime is of the greatest importance. There is no substitute for what they, and they alone, can do to make our society a more secure place. The part that companies already play, and their potential to make a far greater contribution, has been strangely overlooked by government. There is a vacuum in public policy that the recommendations in this paper aim to help fill.

2. CORPORATE SOCIAL RESPONSIBILITY (CSR)

Whether and how companies should share in crime prevention is a question of corporate social responsibility (CSR), and we must therefore locate this paper in the wider context of debates on that issue (Joseph, 2001; Joseph & Parkinson, 2001). Almost none of the domestic CSR debate in Britain is about crime and its prevention—and in the rest of this report we make a good deal of the strange blindness of policy to what businesses can do and have done to reduce crime. It is nevertheless likely that the issues that arise in discussing CSR generally will apply equally to the role of companies in reducing crime. Both debates are about why companies should be expected to contribute to solving problems that have not traditionally been considered within their responsibility.

In the literature, the most common way of demarcating the domain of CSR is to say that it covers all those actions by business which serve the public interest, but which are not required by legislation or similar binding rules.[2] So it is not because of CSR that companies give maternity leave, it is a statutory duty. But it is CSR to employ ex-offenders to reduce recidivism, there being no requirement to do so. Both actions can be said to be in the public interest.

That definition draws the line between CSR and the law. Another line distinguishes between CSR and direct economic benefit. CSR is about actions that do not directly serve the main purpose of business: to make profit. To employ ex-offenders as a previously untapped source of labor in times of full employment is not CSR, it is good commercial sense. To adopt that recruitment strategy in order to contribute to the reduction of offending is CSR, it is not done for commercial reasons. Our concern in this paper is with CSR as the middle ground, where the contribution to crime prevention is neither required by law nor of direct benefit to profit.[3] It may, however, quite legitimately be done because it is of indirect benefit to profits, for instance by enhancing the firms' reputation. Indeed, one strategy that government can employ is to introduce measures that effectively manipulate the market so that CSR does enhance the profits of a company that contributes.[4]

But before we go on to explore the potential for such a strategy, it is important that not all decisions by companies are determined by what is profitable or by a legal requirement.

Beyond the Business Case

Some advocates of CSR justify that activity in terms of the long-term profitability of the firm—"the business case." That analysis usually works something like this. It is in the interests of the typical company to access the benefits to be gained from serving the community in which a company operates. This action ranges from involvement in community safety projects to the employment of ex-offenders. Companies do not normally benefit from these schemes directly in terms of short-term profit. Rather, they offer longer-term reputational and competitive gains, such as:

- enhanced company image (easier to recruit, edge over competitors);
- preferential purchasing by socially aware consumers;
- employee satisfaction resulting from community engagement;
- less targeting of businesses as crime rates fall generally; and,
- a safer and more prosperous community that can create higher demand.

A frequent fault of these arguments is illustrated by the last item of the list, which includes the (true) idea that prosperity benefits business profits generally with the (false) conclusion that it therefore pays the "typical company" to reduce crime regardless of what everyone else does: the free-rider problem.[5]

But the unifying idea is that these reasons are essentially instrumental: it will pay you in the end if you do this. And there are certainly cases when they are powerful: Body Shop (a retailer of health and beauty products) and the environment is an example. Governments and pressure groups are right to emphasize them, because the ideology of stock market-centered business is heavily committed to the idea of hard-headedness and profit, and any respectable chief executive needs to conform to that ideology.

However, companies and chief executives, whatever they may say, have characters and motives more rich and mixed than the impoverished business case allows for. This is even truer of employees other than those at the top. If we are to analyze the reasons for companies to take responsibility for reducing crime, we need a more realistic account than the reductionist assumption of a single principle—that it pays them if only they knew it.

The fault of limiting the analysis to that reason alone is not only that it is often false, but also that it fails to connect with the much richer variety of motivations that can be harnessed to change corporate behavior. Put simply, the business case will not deliver enough. It is particularly weak in the case of designing out crime, because it may actually pay companies to produce products that are stolen because they will be replaced with a new purchase.

Legal Requirement or Voluntary Contribution?

One option is to say that the state should decide what rules should govern the public interest actions of companies and then pass appropriate legislation (or ensure otherwise that such behavior is enforced, perhaps via self-regulation) when those actions do not follow straightforwardly from the pursuit of profit.

There is, for example, a long established and effective system in England of measures that penalize (whether via civil liability or criminal prosecution) companies for operating dangerous fork lift trucks, designing cars with faulty brakes or selling toys with lead paint. Very often, such remedies or penalties are based on specific regulations. Alternatively, they rely on a general duty of care not, by negligence, to cause damage to the user of the product or service. If Parliament thinks it right and fair and effective to pass legislation to make companies conform with their responsibilities, then it can do so, cutting short a debate on whether companies should or should not comply. Thus, there is no need to discuss how companies are motivated to promote health and safety at work, if there is clear and workable legislation enforced by an adequate inspectorate and penalties.

But in the case of crime, we are a long way from having clear principles that can be embodied in law, and it may be that, in this case, it will not be possible to do so. In that case, companies have to have reason to cooperate or the improvement will not take place. For example, we want goods and services to be designed in a way that does not present an opportunity for crime. (Where this has not been done in advance, perhaps because the type of crime was not anticipated, retrofitting may be possible: but that is second best.) Legislation is poorly placed to make this happen. The government generally does not know until they appear what products and services, with what crime potential, are in the pipeline. The companies, and the companies only, may be able to anticipate what may cause crime and how that might be prevented. Unless they more or less volunteer to take the lead, the initiative will not be taken.

Legislation works well when there is general agreement that the behavior to be forbidden is wrong, or that a well-defined, positive duty to act is required. It must be quite clear what that behavior or requirement is, and hence it must be possible to define it sufficiently clearly for the draftsmen to do their work and for judges to enforce the law. At that stage, most companies will want legislation to catch the "cowboys" who have not signed up for the voluntary rules: it is hard to accept that you should spend money to behave well if others are allowed to free ride. If these conditions are met, legislation has the merit that everyone is caught by it, not just the company whose products and services are in the public eye. There is clarity, the minimum of discretionary intervention by politicians, and an open adjudication process.

Voluntarism works better when these conditions, particularly those of clarity, are not fulfilled. Industry has the skills to say what is practical, what is not. It knows better what the problems are and how they are to be solved. Statements of principle work well when conditions, particularly technical conditions, are changing unpredictably, and flexibility is needed. The law works badly, unless you are willing to allow vague drafting and much legal or administrative discretion.

It is clear to us that at present most of what needs to be done in this field is to do with voluntarism rather than the law. We see the way forward as one of partnership and shared responsibility, above all because this area of policy is untried and we need everyone to help work out what it is best to do. That attitude of co-operation is not best fostered by legislation. Voluntarism includes what is now known as soft regulation—meaning measures other than legal requirements that motivate companies to act in the public interest—such as guidelines and non-mandatory codes or practice, "naming and shaming," and public monitoring of performance.

Company Motivation on Crime

What companies can, will and should do cannot be reduced solely to considerations of profit or the law. Thus our account focuses on the wider variety of reasons for acting that operate on companies, a much fuller menu than the reductionists allow. The analysis is about mixed motives. It is not about overarching requirements such as profit maximization, or the law, or indeed generalized notions of corporate responsibility. It is the acceptance that there are a wide variety of reasons for companies to act on crime, and for expecting them to do so.

We can distinguish the following seven ways in which companies contribute to crime prevention, highlighting different motivations and obligations. The list therefore helps to clarify which are the crimes in the middle ground on which our analysis focuses.

1. *Assistance to the criminal justice agencies.* Even on the traditional view of the division of labor, in which responsibility for crime is largely left to the State, companies have an obligation to assist in clearing up crimes—by giving evidence, for instance. These are no more than the obligation that any citizen has, and should be uncontroversial. As with individuals, it is not clear what are the limits to such obligations. But that does not undermine the notion that such obligations exist.

2. *Compliance with the law.* However devoted to the interests of shareholders, companies should conform with the law—for instance, compliance with health and safety regulations. That is not because it may cost them money if they do not. It is because, like any citizen, they have a duty so to do, and that duty exists whether or not they benefit from so doing. The attraction of the law as a way of getting companies to fulfill their responsibilities is that it cuts through questions about why companies should cooperate. Thus companies take responsibility for health and safety as a result of legislation, and there is no discussion about whether they should do so, because they have a duty to conform with the law. It leaves open what the reasons the state has for creating such obligations.

3. *Crimes against business by its employees.* In contrast, crimes by the company's own staff, such as theft of stock, fraud, or stealing the petty

cash, have an immediate impact on the business itself—it loses profits. So there are good incentives for the individual business to take the lead in dealing with such insider crime. Similarly, offences such as stealing the contents of handbags from offices, although not directly bad for profits, are of plain and direct concern to the business and it has an incentive to deal with them. Certainly, other agencies are needed too—the police to apprehend the offender, the courts to try the case. And such crimes may have wider implications for society—petty office theft can be the way that persistent offenders start work, and such crime is part of a general pattern of lawlessness and indiscipline which disfigures society. So we all have an interest in dealing with it. But there is no doubt that the affected business is in the front line, and that the direct incentive to do something about it is internal to that particular, individual, business. This is a case of self-interest.

4. *Crimes against business by outsiders*. A close neighbor is the crime against the company carried out by outsiders. The sneak thief who comes in to steal laptops from desks, the shoplifter—these are crimes which affect profits or at least the operation of the business directly and the company has a clear interest in them. The difference is that the company is likely to need help not only from the criminal justice agencies but also from its neighbors. So this is not just that in the traditional way the police should be notified of the crime and helped to pursue it. Rather, if shoplifting is to be reduced, there may be a limit to what the company can do inside the store. What may be needed is a shared effort of some kind, shared between the police, the store, the local authority, the landlord, to prevent shoplifting not only in that one shop. So partnership is needed, but the motivation of the company remains much the same—it wants to stop crimes like this because they are crimes against its own interests.

5. *Theft or misuse of a product.* A decisive step away from the previous example is the class of crime with which this paper is most concerned. When in the 1980s joy riding became commonplace, the offences were not against the manufacturer, but against the owner whose car was stolen or damaged: the other driver in the crash; the shop that was ram raided; the victim of the accident. Joy riding was made easier because cars were insecure and hence easy to steal (this was before the introduction of immobilizers and similar security measures—on which we have more to say below). But the crime was not against the

manufacturer. So a manufacturer whose vehicles were easy to steal would only be directly concerned with that crime if it affected the willingness of customers to buy from him. The offence as such is of no direct concern. So he has no direct incentive to redesign the car—unlike the retailer who will redesign the store layout to prevent the crime of shoplifting. And you could say that the state should take prime responsibility for crime reduction in this instance, by improving police patrols and crime prevention.

6. *Abuse of a service.* Similar to the preceding category is the misuse of a company's products or services, as when drunkenness leads to crimes of theft, criminal damage or violence, drug traffickers launder their proceeds through high street banks, or child pornographers use the services of an internet service provider (ISP). These crimes are often neither against the business itself—the drinks company (if the offenders were causing damage in the bar they would fall under three or four above), the bank or ISP: nor against the customer—indeed the customer commits them: but against third parties. The service provider may—or may not—be in a strong position to prevent it.

7. *Crime reduction unconnected to the company.* Then there are cases where businesses can indeed contribute something to crime reduction, but it is hard to argue that any particular business has any strong, individual, direct responsibility to do so. For example, it would undoubtedly reduce re-offending if companies were more ready to re-employ offenders—and there are companies that run recruitment and training schemes to do just that (an issue we explore further below). And it may be right, as a matter of public policy, to introduce legislation or other rules to make this happen (as indeed, in the case of ex-offenders, is required in Canada). But the principle applies to all companies, and makes no claim that any particular company has, resulting from its own position, and any direct responsibility for re-offending. A company that runs a program to rehabilitate offenders does not do so because it has any direct responsibility for having created the crime, and its only interest may be the same as the rest of us, that crime be reduced. There may be reputational advantages in terms of its image with customers, employees and government. But there is no direct responsibility on that particular company to take that action.

Our primary interest is in categories five and six—theft or misuse of a product or service, because of the clear responsibility companies have to act in these cases. The final example, crime reduction initiatives with no direct connection to the company, is nevertheless a fine example of CSR and any initiative to develop corporate responsibility in this area should encourage it.

Basis of Responsibility

What are the defining characteristics of the crimes in categories five and six? What gives purchase to the idea that companies should in these cases take responsibility for helping to reduce crime?

First, the company, its products or services, is an important part of the causal nexus that produces the crime. That does not mean, for instance, that alcohol is the sole cause of disorderly behavior—individual choice, peer pressure, despair, many other things are among the causes. But what the company does is part of it.

Second, the company has a responsibility if it can do something to reduce the crime. That is not always the case. Kitchen knives are used to commit murder, so the knife retailer meets test one. But he can do nothing to stop murders short of not selling such knives—there is no practical way of distinguishing between cooks and assassins at the point of sale.

Category seven does not meet these tests. It would certainly be in the public interest if companies were to concern themselves with crime in that way, but it cannot be said that it is a particular company's responsibility to do so. And neither the law nor profit is sufficient motivation for them to act. Encouraging such contributions is important. But our primary focus is on cases where companies' products and services are at the heart of the problem, and they both can—and have a responsibility to—do something about it.

The case for corporate responsibility is stronger the greater the role that business products and services play in the causal nexus, compared with other factors: and the greater the role which the business is capable of playing in reducing crime, compared with other players. But it is clear that there are limits, albeit not easily identified, to the responsibility which companies can be expected to take even when these two tests are met. The case of the kitchen knife exemplifies the problem of basing responsibility solely on the criterion that the company is central to the design and production of the criminogenic product. Nobody would say that knife companies are responsible for stabbings.

Two difficult examples deserve a closer look. The case of clandestine entrants into the United Kingdom illustrates well the complexities of deciding what responsibilities companies have for solving problems which fall outside their commercial interests, and what legal burdens the state can legitimately impose on them to achieve its or society's purposes. Conveniently, a recent court case provides considerable insight into the detail of the issues involved.[6]

In the late 1990s, the problem of clandestine entrants was a major policy issue. A clandestine entrant is typically one who hides himself in a cross-channel truck, gets into Britain, say at Dover, and then claims asylum.[7] In 1999, the British government introduced legislation which imposed fines on truckers if clandestine entrants were found in their trucks on arrival in Britain. The fine is £2,000 per clandestine. Some fines have been very considerable. It is no defense to say that you did not know that you were carrying the entrant. You are required to have carried out, as necessary, repeated and detailed checks on the integrity and security of your truck against such misuse, in accordance with a code of practice and your own codified procedures. Only if you can show rigorous conformity with these rules may it be possible to escape the penalty.

It is agreed, at least by British public opinion and its government, that reducing clandestine entry is highly desirable. It is also plain that if truckers succeeded in finding and excluding all such entrants, the problem would be much reduced. So for those who wish to deal with corporate responsibility for crime reduction by asking no more than whether the services provided by business provide the opportunity for the offence, and whether they could do more to stop it, the answer is clear. It is the truckers' trucks which create the opportunity, and they can solve the problem by eliminating the opportunity.

But this case shows the problems of that simple, delivery based, "do whatever is necessary" approach.

1. How much are the truckers meant to do? How often should they search, at what cost to them? The Code of Conduct is long, complex, and onerous.[8] What is reasonable? There is surely some limit to what we are entitled to impose on them. Where is that limit?

2. How big should the fines be? As with other penalties, they must fit the crime—transportation for sheep stealing would have been going too far even if it had worked as a deterrent. £10 may be acceptable for not having a tube ticket. But the average fine in these cases of £12,000 could ruin a sole trader with one truck.[9] There must be proportionality.

3. The law makes the driver liable, even if he has carried out the checks which his owner requires, if that system of checks is not an effective system. And the employer who has put in a good system may be liable if his employee fails, despite training and instruction, to carry out the checks properly. Is that fair? More generally, how far should we go in making a person responsible not only for his own acts, but for acts of others over whom he cannot exercise any effective control?[10]

4. What rights does the truck driver have? Can he be detained? Has he got the right to silence? What is the onus of proof? It may be sensible to skate over such issues in the case of a £10 fine. But what if it is £12,000?[11]

5. The reason that the U.K. government needs these rules may not be simply to stop entry. It may have sufficient powers for that under the 1971 Act. The real purpose may be to avoid the cost and inconvenience of dealing with people who once in Britain are able to claim asylum.[12] Why should the trucker be required to help with that objective?

6. More generally, the problem of refugees, asylum and illegal entry results from a complex set of political, legal, and economic factors far more powerful and fundamental than whether cross channel trucks are available. It is not clear what share of the burden of solving the problem should be imposed on truckers as compared with, for example, the relevant governments. What responsibility should be placed on the French police for using a refugee centre at Sangatte which facilitates multiple attempts to stow away?

We provide this list, and go into such detail, to emphasize that what needs to be done depends on the circumstances of each case.[13] There are limits to what can be achieved in the abstract, apart from context, by general principles. Particularly, general statements about Corporate Social Responsibility can only go so far. The questions outlined above should not be dismissed as peculiar to this case, nor merely technical, nor only for the lawyers. Saying what companies can and should be expected to do will depend on among other things the facts of the case. Those facts are what will enable us to say in each case what is reasonable or fair or proportionate. Such judgments are inescapable.

Another difficult area which has caused much controversy during the winter of 2001/2002 is the responsibility of the mobile phone operators for reducing the street theft of mobile phones. Such thefts are seen as having reached epidemic proportions, and some research shows that as much as 28% of robberies are associated with these products. The motives of the, often young, offenders can be complex. It may not be that they simply want to steal a usable phone; possession of a glamorous object may be enough, and such thefts may even be a way of disciplining or bullying outsiders who stray onto their turf (Harrington & Mayhew, 2002). But surely one way to make the theft less attractive is to make the phone unusable when stolen. The newer networks—Virgin, T Mobile (previously One2One), and Orange—have the technology to disable remotely both the Subscriber Identity Module (SIM) card and the handset itself via its International Mobile Equipment Identity (IMEI), a unique number given to every cell phone. The older networks—MmO2 (previously Cellnet), Vodaphone—have older software, which does not allow that to happen. The debate and indeed the dispute between the government and the industry centered on whether these companies should make the technical changes needed. On the one hand, the crimes plainly arose from an opportunity associated with the companies' products: and they could, technically, do something about it. On the other, changing the software costs a lot of money. How much can the companies reasonably be asked to spend to solve a crime problem? Surely that is a matter for the state. It is not the fault of the companies that mugging and street crime continues to flourish. Should not the state solve the problem by better policing?

In the event, the older companies agreed to enable Generation 2 mobile phones to be call-disabled if stolen. But here again, there are no clear-cut principles, which tell us what is fair, or reasonable, or proportionate. It is a matter of the facts of the case.

The example of mobile phones can also illustrate how easy it is to lay everything at the companies' door. Although the new disabling procedures will certainly help, nobody believes this will be the end of such street crimes. Thieves have a variety of motives, as we say above, beyond just wanting to have a working phone. Any disabling procedure requires that the theft be reported. But it is unlikely that there will ever be 100% reporting, particularly of thefts of pay-as-you-go phones. Similarly, the use of statistics, such as that 28% of robberies are associated with these products, conceals possible inaccuracies in data collection and does not show that, for example, in the case of a mugging the mobile phone was the target rather than cash or credit cards.

Before crimes associated with products or services are tackled by the weight of public, political, and media attention, they have to become salient.

Salience is often associated with alarm and with over-attribution—of the importance of the product or service in the generation of the crime, and of the magic bullet results to be expected if only business were to cooperate. Things are rarely so simple.

A useful comparison is with the area of product liability. Stringent laws exist (both U.K. and European) to make a manufacturer liable if they produce a product which is defective and causes physical injury to the consumer as a result. A defect can relate to an inherent flaw in design, or to the presence of dangerous (for example, flammable) materials in the product. European law has now been tightened in this area so that in a court case a producer is liable (and can be fined) if a product is merely defective: it is no longer necessary for those suing the company to prove that the customer did not misuse the product or was negligent with it.

Factors That Influence Motivation

We know from wider CSR studies, unconnected with crime, that there are a number of reasons why small, medium and large companies—despite their very differing levels of resources—may choose to act in a socially responsible way; some to the extent of having a thought-through CSR agenda. We examined these to see which of these operate in the case of crime, which do not, and whether they all could do so.

There are a variety of pressures and considerations that influence the environment in which companies act and what they do. Business people, like everyone, have mixed motives and, as we have argued, it would be a mistake to think that they act ruthlessly only on what directly leads to profit.

- *Pressure groups, therefore, do matter.* The importance of environmental issues (low-energy washing machines), fair trading (Latin American coffee), additive-free food, and so on, has been highlighted by pressure groups, which have raised the consciousness of companies, consumers and government. This effect is not quite the same as the pressure of the market. It is more general and operates whether or not it can be tracked to identifiable effects on sales.

- *Peer pressure.* Like any other group, people in business and industry like to be respected by their peers. The role of Business in the Community is important in that context. Led by senior practicing business people, it identifies issues such as the stability of communities, where companies have a contribution to make, and where coordinated recognition of that contribution sets an example to all businesses of what can and should be done.

- *Reputation and self-esteem.* Business people want to be well thought of not only by peers but also by customers and others in a position of influence, not least the government. This is true of the company chairman looking for a knighthood, and of the staff who want to be proud of what they produce—and they readily accommodate the idea that they can indeed be proud because the product is safe, or additive free, or reduces crime.
- The press, particularly the *specialist press. What Car?*, a British consumer magazine offering independent reviews of new cars (www.whatcar.co.uk), was important in establishing "steal-ability" as a criterion of what constitutes a bad car. Again, this may or may not be directly or at once reflected in what people buy. But it affects crucially the climate in which businesses make their decisions.
- *Pressure from investors.* The Association of British Insurers, for example, has introduced new guidelines (Association of British Insurers, 2001) to manage the risks to shareholder value arising from social, ethical and environmental factors, and the Turnbull Report highlights the duty of companies to recognize and manage the risks arising from "legal, health, safety and environmental, reputation and business probity issues" (ICAEW, 1999). These guidelines and duties arm investors such as Friends Ivory and Sime to put pressure directly on companies to take such issues into account (Friends, 2001).
- *Fear of regulation.* Companies can be motivated by generalized fears that if they do not behave well, the government will take action. The fear may be of new legislative requirements or constraints, or that they will not get the necessary license or contract. Again, this can be tracked back to a fear that government intervention will impact on profits. A better way of understanding it is that it is better to have the government on your side than against you, regardless of exactly what it is that it might do.
- *The desire to be a good citizen.* Business people live in the world, and they want to be part of it like others. On the whole, they want to share

the values of other people whom they respect and like. So, given choice, they will do right rather than wrong, at least as much as other people. Not for the first time, (the Victorian era was a more powerful example, fuelled by Christianity rather than by secular social responsibility), business people feel a need to take some responsibility to improve society where they can, and where it is reasonable with regard to their other goals. Corporate behaviour cannot be reduced to obedience to law or profit, any more than that individual human behavior can be reduced to utility or obligation.

- *Profit and legal requirement.* Finally, we must not forget the law and profit. Because we are dealing with CSR, these are not by our definition sufficient on their own in the cases which we are considering to produce action. They operate as limiting cases in two ways. First, by definition. As we have discussed it, CSR excludes the cases where the law determines the action and those where profit does. Second, if consumers change their attitudes, what was a matter of CSR (for example, to make cars safe) becomes a matter of profit: consumers want and will prefer to buy safe cars. As opinion consolidates, what was a matter of consumer and company choice—whether to sell or buy a secure car—became a matter of EU law: immobilizers must be fitted. But even before we reach one or both of these termini, when we are in CSR territory, both profit and law may play a part. Indicative codes of conduct may matter for what companies do as a sort of precursor of the law. Some customers may want crime-reducing products, which nudges companies in the direction of such products becoming standard. But in CSR territory those reasons are not enough on their own.

If products and services and goods are to be designed to prevent crime, only companies can do it. If they do not do that particular job, the police cannot, the customer cannot, and nor can the government. Companies alone know what to do to design out crime and have the direct power to do it. This is the reason for them to take on the responsibility and for us to expect them to do so. It is not a conclusive reason—no single reason is. And it does not mean that every company has at once, regardless of cost, to do everything possible to design out crime. Responsibilities are rarely absolute or unqualified. Reasonableness and proportionality comes in here as everywhere, and the contribution of others is indispensable.

3. PUBLIC POLICY AND OPERATIONAL INITIATIVES

Until recently, there has been little public policy on the participation of companies in crime prevention. If there has been an attitude, it may be fair to describe it as traditional with opportunistic episodes. The traditional element is the idea that companies, broadly, have no substantial role to play. They are consumers of the services of the criminal justice system, and have no major actual or potential responsibility for crime prevention or any other such initiatives.

There are exceptions to this. The licensing system for the sale of alcohol and entertainment providers, for instance, imposes legal constraints and obligations on those service providers designed to limit the potential for crime and disorder, among other objectives.

The norm is ad hoc and opportunistic. From time to time crimes related to particular products and services rise to the top of the public and political consciousness. Something has to be done fast, and the relevant businesses are made a central part of the solution. The recent examples include the dramatic rise in theft of mobile phones, leading to Ministers' exhortations to operators to ensure stolen phones are unusable.[14] The classic case is joyriding in the 1980s, when Mrs. Thatcher, it is said, got the car manufacturers into a room and told them to do something about it; which they did. Similarly, the appeal of alcopops (soft drinks with a low alcohol content) to under-age drinkers in the 1990s led to action by the drink producers as a result of government and media pressure. The companies established a voluntary code of practice to control the naming, packaging and merchandising of these drinks, to counter criticism that these products were deliberately targeted at teenagers.

These ad hoc solutions may resolve an immediate problem. But they take place after the event, and could have been more effective if crime prevention had been a priority during the design stage. What industry has done in some sectors over the last 20 years has been impressive and effective. It now needs to be mainstreamed. We need to think more about why businesses should be responsible for crime prevention and how; about what kind of co-operation between businesses and the other players is needed, for what and where; and about what is the best way to get business more involved, and what are the most appropriate motivating forces on the spectrum from legislation to self-interest.

If we look at the public policy background over the last few years, it is best to start with the government's Crime Reduction Strategy published in November 1999 (Home Office, 1999a). This covers a wide variety of initiatives, from helping victims to young offenders. Of direct relevance to this paper are the steps taken to tackle vehicle and property crime, and the Crime and Disorder Reduction Partnerships.

Vehicle and Property Crime

The Vehicle Crime Reduction Action Team (VCRAT), chaired by Mike Wear of Ford, was set up in September 1998 and worked to a target of a 30% reduction in theft of and from vehicles over the five years to March 2004. The team was set up and staffed by the Home Office, and its membership included representatives from the police, business and industry. In September 1999 it published *Tackling Vehicle Crime—A Five Year Strategy* (Home Office, 1999b). Its main recommendations were:

- improved security for both old and new cars;
- better policing and community response to target prolific offenders, crime hotspots, and the market for stolen goods;
- improved car park security; and
- new procedures to reduce the market for stolen vehicles, for example by stopping stolen vehicles being given false identities.

In October 2001, a progress report issued by VCRAT looked at how well its recommendations were being implemented across industry and by the police, and it listed substantial progress on the 22 specific recommendations in the 1999 report (Home Office, 2001c). The 2001 Vehicle (Crime) Act has particularly helped the control of license plates. A number of measures are coming into force this year, such as the establishment of direct link between the Driving and Vehicle Licensing Authority (DVLA) and the Motor Insurance Anti-Fraud and Theft Register (MIAFTR).

Similarly, the Property Crime Action Reduction Team (PCRAT) was set up by the Home Office in November 1999 to look at ways of reducing crime against property, both residential and commercial. Its initial report, *Tackling Property Crime*, was published in May 2001. It focuses on recommendations which include: encouraging private builders to improve the physical security of buildings and surrounding environment; publicizing the benefits of improved security so that the public will demand all buildings to have adequate security measures; reducing the value of stolen property to the thief; and preventing arson. The report (Home Office, date unknown) also points to the need for long term funding to entrench the long-term crime reduction strategy.

Crime and Disorder Reduction Partnerships (CDRPs)

The reference in the Strategy to Crime and Disorder Partnerships stems from the statutory duties imposed by the Crime and Disorder Act 1998 on local authorities and the police to cooperate in the development and implementation of a strategy for tackling crime and disorder in their area, and to consult with a wide variety of local players, including health authorities and probation committees.

These provisions have been implemented, following the publication in July 1998 of *Guidance on Statutory Crime and Disorder Reduction Partnerships*, and by 2001 there were some 400 partnerships based on these provisions. Their scope is much wider than the concentration on the role of the police and local authorities in the strategy suggests. The 1998 Act has been the catalyst for a variety of local partnerships against crime, in some of which the role of the police has not been central but in which business has played a part. There is as yet no publicly available survey of the overall effect or success of these partnerships.

The Retail Crime Reduction Action Team has produced six guidance documents for retail partnerships. It has developed the Safer Shopping Award Scheme, promoted the Business Information Crime System, and registered increased numbers of retail crime partnerships.

Other Initiatives

As well as these initiatives, there have been a number of analogous moves, not necessarily foreshadowed by the Crime Reduction Strategy or based on the 1998 Act.

It is not necessary to have a formally constituted Task Force to achieve progress through consultation and partnership. The August 2000 Action Plan for Tackling Alcohol Related Crime, Disorder and Nuisance promotes the objectives of reducing the problems arising from underage drinking, reducing public drunkenness, and preventing alcohol-related violence, using both existing legislative provisions (for example, a prohibition on sales to under-18s) and voluntary procedures of the kind which we examine in the Drink Industry case study (#2) below. This Action Plan was the result of consultation with the industry, including the Portman

Group self-regulatory organization. There is as yet no publicly available assessment of its success.

Similarly, there have been extensive consultations over the years between the Home Office and the credit card industry on crime and its prevention. These consultations are ad hoc and have not required the creation of an Action Team, maybe because the industry is well organized already through the Association for Payment Clearing Services (APACS), which represents almost 100% of the relevant companies. And the resolution in February 2002 of the problem of disabling stolen mobile phones resulted from prolonged discussions between government and the industry.

The Foresight Crime Prevention Panel, part of the wider Futures initiative of the DTI, started work in April 1999 under the chairmanship of Colin Sharman, previously senior partner of the accounting firm KPMG. It concentrated on the future of crime and, in particular, on the role of science and technology in reducing crime. It published, in December 2000, *Turning the Corner*, which includes recommendations on mainstreaming crime reduction in the decisions of government and business thinking, and addressing crime at all stages of a product's life cycle (DTI, 2000).

This panel has operated as a focus and forum for the discussion of such topics by police, companies, officials and academics. Its Products and Crime Task Force considered the scope for developing a code of practice on product security for electronic goods and considered the incentives needed for companies to take up these measures. It also looked at potential new crime threats, such as that of e-tailing, the delivery of retail goods sold via the Internet, which can provide dishonest delivery staff with information on when customers will, and will not, be at home. The Foresight Crime Prevention Panel closed at the end of March 2002.

The Forecourt Watch scheme is being pioneered by Thames Valley Police and the British Oil Security Syndicate (BOSS) in an effort to reduce the number of people driving off without paying for fuel. The scheme operates by putting in place technology that scans vehicle registration plates to identify stolen cars, or identify known offenders. It is currently being tested by police and petrol retailers in Milton Keynes (*Communicate*, 2001).

A full list of the initiatives taken by government and public agencies over the last few years to combat crime with the participation of business would include: the Chipping of Goods Initiative; the production of crime reduction tool kits; the Designing Out Crime Unit in the Home Office; the provision of £15 million to small retailers for crime reduction in high crime areas; and the Vehicle (Crime) Act 2001. An excellent source of information is *Communicate*, the magazine of Business Crime Check, which is distributed to 7,500 public and business subscribers and reports quarterly on a wide variety of initiatives. Although the list of initiatives is impressive in its range (and perhaps, though evaluation is rare, in its achievements or promise), there is surprisingly little recognition of the importance of such partnerships. Yet all the above depend heavily on cooperation.

In December 2001, the Business Crime Survey published by the British Chambers of Commerce,[15] based on responses from 2,914 of their members, found that 34% of businesses had never had any crime reduction advice and 82% were not aware of a community safety partnership in their area.[16]

Government Policy

Recognition within government has been slow to come. You look in vain for any mention of business in the relevant sections of the 1998 Crime and Disorder Act—and will find it only in the published guidance, where "employers"—a quaint term in this context—appear as number 19 on the list of bodies which local authorities and police should consult. The Home Office White Paper *Criminal Justice: The Way Ahead*, published in February 2001, makes little mention of the role of companies, except for reproducing the five recommendations of the Foresight Crime Reduction Panel (2001, p. 130) in an appendix. Its introduction refers to Crime and Disorder Reduction Partnerships and mentions the role of the police, local councils, and others, but not explicitly business (Foresight Crime Reduction Panel, 2001, p. 5). Similarly, the DTI's *Business and Society, Developing Corporate Responsibility in the UK* contains little on crime (DTI, 2001).

The most recent White Paper, *Policing a New Century: A Blueprint for Reform* marks a step forward (Home Office, 2001a). It includes a section entitled private sector responsibilities, which refers to the contribution made by the Vehicle and Retail Crime Reduction Teams and says that: "The Government is determined to use to the full the ideas and professional expertise of the private sector in addressing problems which affect us all." The White Paper focuses on the "principle" that the policing requirements on private property (such as football stadiums and private shopping malls) are paid for by the venture concerned, and showcases various schemes where business has paid for additional policing services. Noting that excessive costs of policing events can still fall on the police, however, it says that it intends to explore with the drinks industry and licensing authorities (Home Office, 2001f, para. 5.48):

> the potential for building on and extending voluntary agreements already reached in a number of areas which have resulted in entertainment venues making a contribution to the policing and public order costs generated by their activities.

Referring to other kinds of business activity that can generate opportunities for crime, such as the theft of mobile phones and credit card fraud, and the joint initiatives that have been taken to address it, the White Paper concludes, with no detail attached (Home Office, 2001f, para. 5.50):

> These projects have developed on an ad hoc basis to date. In the future, we intend to develop a clear framework for public-private partnerships to tackle specific business related crime.

The recommendations in this paper suggest how this might be done.

Non-Governmental Initiatives

Outside government, and of great importance in the current debate is the Design Council's report by Ken Pease (2001), *Cracking Crime through Design*. Pease argues that it is vitally important to win the argument that changing design will reduce crime and that significant displacement to other kinds of offences will not occur to negate this. The evolving nature of crime and crime prevention methods is important too. In the past, the report argues, the typical relationship between product innovation and crime has been as follows: Innovation without consideration of consequences—criminals reap the harvest, the crime prevention solution is retrofitted. Pease argues that it is necessary to "intervene" by considering design at the innovation stage and not at the end when the crime wave has already occurred, as is usually the case. Progress will only come if there is consensus on this approach "at the top" and the DTI and Home Office work together to bring industry on board. His recommendations include increasing incentives for companies through approaches such as "naming and shaming" and awards for the best crime-resistant products.

4. CASE STUDIES

The case studies that follow are not designed to provide a full account of all the crime prevention initiatives taken in these industries over the last rpose is rather to look at some particular developments, ..ıs about how industry and its partners work together, what ..ions they may have, and the role of government and the criminal ,.ıstice agencies. These lessons are then brought together in the following section, leading into our policy recommendations.

1. Cars

The Problem

The joyriding phenomenon of the early 1980s was not only dangerous for the youngsters involved, but innocent people were killed in the resulting accidents. More generally, the problem was the rising levels of car crime. Total vehicle crime almost trebled from over half a million recorded incidents in 1979 to 1.5 million by 1990.

The Solution

Particularly through the fitting of immobilizers, the car industry transformed the design of its products to reduce theft.

The Story

It is natural that vehicle crime should have salience.

1. The size of the problem. Vehicle crime accounts for roughly one quarter of all recorded crime in the U.K. (Home Office, 1999b, p. 6). There has thus been pressure not only from within the industry but also from Government, the police and consumers to address the problem.
2. Cars are one of the most ubiquitous consumer "goods" on the market. The high level of car ownership guarantees that the issue of car security is something that everyone has an interest in.
3. Cars are expensive pieces of equipment and attractive to the thief. Their cost makes it worth investing time and energy into developing crime prevention measures.

The steps taken to reduce vehicle crime included:

- new design features;

- new systems of vehicle and vehicle component identification and registration;
- changes to the after-sales market; and,
- changes to the environment in which cars were used.

Design features introduced by manufacturers to improve security include car alarms, laminated glazing for windows, codes for car radios, electronic immobilizers and enhanced door-locking systems such as deadlocks. Immobilizers are widely regarded as the most effective crime prevention tool in cars to date and "doubtless helped in achieving the 26% reduction in vehicle crime that [took] place over the 5 years ending March 1998" (Home Office, 1999b, p. 8).

Of the 1 million plus vehicles sold in 1998 by the top five manufacturers and fitted with Thatcham compliant immobilizers, fewer than 0.25% had been stolen by the end of the year (Home Office, 1999b, p. 10). Immobilizers arguably had most success in deterring opportunistic theft (whereas the professional has the expertise to by-pass the technology). The Vehicle Crime Reduction Action Team (VCRAT) believes that the impact of the EU's Directive, requiring all new cars to have immobilizers fitted from 1998, will be felt over the next few years as older vehicles are replaced. It estimates that this will prevent about 120,000 offences in the U.K. between 1999 and 2004 (Home Office, 1999b, p. 8). VCRAT has also proposed an extension of EU legislation to cover deadlocks in cars. This it believes could significantly reduce "thefts of" and "thefts from" a vehicle because forcing door locks is one of the typical entry methods used by the criminal. At present only 30% of new cars have deadlocks fitted as standard and the Home Office is endeavoring to persuade manufacturers to cooperate (Home Office, 1999b, p. 9).

Changes to systems of vehicle identification and registration have also been important in rendering more difficult the sale and use of stolen vehicles. It is now an industry standard to have not only the engine but also other component parts marked with the 17-digit Vehicle Identification Number (VIN) and to display a visible VIN on the car windscreen. However, some within the industry are keen to see this backed up by legislation to prevent any tampering with the VIN. This is not an offence and it allows the criminal to obscure a number to cover up a stolen part or engine.[17] It has also been suggested that the VIN be placed on a random selection of components in order to make it harder for criminals trying to sell on vehicle parts. Another change due to come into effect in June

2003 was the computerization of MOT records. (Motorists seeking to renew their vehicle license must produce a current MOT certificate for all vehicles over three years old. Called MOT because it was introduced by the then Ministry of Transport, this certificate is issued by an authorized agency following a comprehensive check on the road-worthiness of the vehicle. Computerization will make it harder to forge MOT documentation as there will be a national on-line database to back up written records.)

In the after-sales market there has been an attempt to establish standards for vehicle security products sold to customers. One standards body is “Sold Secure’, a division of the Master Locksmiths Association that tests security devices for a wide range of vehicle applications. Their advice is to customers is to purchase only those products which have been shown to be capable of withstanding attack and are labeled as such (Home Office, 1999b, p. 13).

There is still room for improvement. One difficult question is how to encourage the owners of older cars to have immobilizers retro-fitted in their cars. VCRAT explored this issue in a recent publication (Home Office, 1999b), and concluded that voluntary fitting schemes (encouraged through publicity and other incentives) were not enough.[18] Instead the law should be changed to require cars in certain registration years—it suggested 7–10 year old cars registered between 1991–94—to be fitted with an electronic immobilizer (Home Office, 1999b, p. 9). This would be enforced through the MOT and would address directly the problem of the greater vulnerability of older cars (Home Office, 1999b, p. 9). But the government has not accepted this recommendation, favoring incentives and publicity to bring the change about.

Further progress has been made in reducing vehicle crime by making the locations in which cars are parked more secure. One particular trouble spot has been car parks: the 1998 British Crime Survey showed that 22% of vehicle crime took place here. Secured Car Park schemes have been set up by the Association of Chief Police Officers (ACPO) in an attempt to address this problem and these have proved largely successful—an average reduction in car crime of 70% was observed in sample surveys (Home Office, 1999b, p. 18). By the end of 2001 the scheme covered 900 car parks,[19] with measures including £50 million spent on CCTV from the Capital Modernization Fund.

The fall in vehicle crime in recent years shows that the problem has been substantially reduced.[20] By the end of the 1990s, total vehicle crime had been reduced to the levels of the late 1980s—1 million recorded incidents (Home Office, 2001b, p. 6). This figure is still high and there is much to do if the official target of reducing vehicle crime by 30% in the five-year period up to 2004 is to be met.

Lessons

The immediate impetus came from public and media concern about joyriding in the 1980s, which resulted in a summons by Mrs. Thatcher to the car manufacturers to take action. The action came, initially and above all, from the companies themselves. The motivation of the companies was not primarily commercial. There was no direct market advantage to them at first. But as time passed their reasons for acting multiplied:

- Interest in the issue of car security grew steadily and was reflected in the amount of attention which car magazines like *What Car?* gave to it. Customers wanted to know which car offered the best value for money and vehicle insurance was one of the factors that they considered. Increased car crime made it good sense to purchase a car that was harder to steal/break into.
- Insurers established their own independent test centre at the long established Thatcham unit in 1993.[21] The "attack-testing" carried out here to check how vulnerable a vehicle was to theft was one of the factors taken into account when deciding on a car's insurance category (on a scale of 1 to 19). It was thus in manufacturers' interests to ensure that their cars were "Thatcham-compliant" since they did not want high insurance premiums to deter customers.
- Central government took the lead in efforts to reduce vehicle crime. Its "Car Theft Index" listed the cars most at risk of being stolen and pushed car companies to consider how to design secure vehicles. A number of Home Office working parties were established which worked with manufacturers to disseminate good practice within the industry. The Vehicle Crime Reduction Action Team (VCRAT) is the latest manifestation of these collaborative efforts.
- The EU's 1995 Anti-Theft Directive on the fitting of electronic immobilizers in cars made them compulsory for any car produced after 1998, and forced manufacturers who had not already done so to start fitting them as standard in their vehicles.

- The companies came to accept that producing good vehicles, always a matter of professional pride in the industry, included an interest in producing crime-free cars.

It was also suggested to us that the police were motivated to take this sort of crime seriously not only because it is indeed serious, widespread, and high profile, but because it is the sort of crime which they like dealing with—personal, soluble, to do with cars and linked with non-vehicle crime such as burglary and drugs. True or not, that idea raises the important issue that it is to some extent a matter of discretion which crimes the police choose to concentrate on.

There were a number of key players other than the companies. At the start, when the manufacturers saw that they had to act they lacked many of the technical and security skills needed. These skills had to be brought in or even developed anew at outside research centers. Thatcham and the insurance industry generally were key. So too the pressure brought by "*What Car?*" But it is a stretch to describe this, or indeed the role of the police and central government, as partnership. Most of what was achieved was at arm's length.

The law played an important supporting role. In the end, when the need for immobilizers was generally recognized, and most new cars had them fitted, the EU regulation on immobilizers was important.

The Vehicle (Crime) Act 2001 has also been significant. Its central aim was the development of new procedures at DVLA, and to secure the cooperation of the motor salvage industry, to stop stolen vehicles being given the identity of other, legitimate, vehicles and to help prevent vehicles from being stolen for spare parts. Reducing the market for stolen vehicles in this way will mean that they have less value to the thief, who will, in turn, have less to gain by stealing them in first place. The legislation is a good example of how the law promotes co-operation between the police and industry to bring about changes that will help cut crime.[22]

Car manufacture is a global business. The U.K. is one of the countries with the highest rates of car crime—other markets like Japan and Korea have much lower rates and do not therefore require similar levels of protection. Central Government played a key role in providing the impetus. Thereafter, the work was done mostly by the companies. Nevertheless, the provision of a forum such as VCRAT and government's persistent focus on crime reduction in general and vehicle crime in particular has been important in keeping up pressure to produce solutions. There has been very little public recognition of the success of these measures. The companies paid almost all the money for the development of the necessary technology and its manufacture.

Deficiencies

It can be argued that the measures in, for example, the 2001 Vehicle (Crime) Act came later than they might. And further legislation may be needed, for instance to require immobilizers in old used cars (rather than merely encouraging owners to fit them); and to outlaw tampering with vehicle identification numbers on car parts. Nevertheless, the partnership on car crime as embodied in VCRAT seems to work very well. We address the particular case of car radios and immobilizer decoding equipment in our recommendations.

2. Alcohol

The Problem

Alcohol misuse, particularly excessive drinking, has always been a source of major social problems—from the old days of gin palaces to the lager louts of today. And the brewers and distillers have always been the object of opprobrium for making money out of misery and addiction. So the perceived problem is much as it has always been: drunkenness leading to crime, and to longer-term problems such as family break up and unemployability. The particular problems of the end of the twentieth century can be seen as contemporary examples of the same issues.

The Solution

The establishment in 1989 by the drinks industry of the Portman Group, a private sector self-regulatory body, to deal with products, labeling or marketing which encourage heavy drinking or drinking by the young.

The Story

The Portman Group exists to promote responsible drinking, by deterring the misuse of alcohol in society, and enforcing good practice on the part of producers, pubs and retailers in the sale and marketing of alcohol. It runs public education campaigns, provides educational material to schools and the media, and runs the under-18 proof of age card scheme. Its 1996 Code of Practice, which was established in response to public and government concerns about the appeal of alcopops to the under-18 age group, lays down guidelines on how alcohol should be labeled, packaged and merchandised. It has the backing of all of the main alcoholic drink producing companies in the U.K., and its signatories include over 100 companies spanning producers, importers, wholesalers and retailers in addition to the Group members. It is a self-regulatory procedure, which depends not on legal penalties but on deterrents such as damaging publicity from upheld complaints, peer pressure, alienation of customers, and bad reputation generally.

The most effective section of the Code regulates how pre-packaged products and services are labeled. It obliges the producer clearly to communicate the alcoholic nature and strength of the product and services to the consumer; to avoid in the product's name any association with violent, aggressive, dangerous or anti-social behavior and any allusion to either illicit drugs or sexual prowess; and not to encourage irresponsible consumption such as binge drinking, drunkenness or drink-driving. The Code also seeks to limit the appeal of drinks to under-18s—which had been the original criticism of alcopops—by restricting the following in the packaging and marketing:

- imagery or allusion to under-18s culture;
- designs or marketing techniques (including endorsements) based on or resembling characters popular in predominantly under-18s culture;
- labels which incorporate images, cartoons or drawings of people who appear under 25 years of age; and,
- artificially bright colors or style of lettering predominately associated with under-18s.

It is specified that in the case of mixer drinks not only must the alcohol type be on the label (for example, vodka and lemonade), but also that it must be prominent in terms of "color, style of lettering and field of vision." And the color or texture of the drink is also mentioned as a factor that can influence the decision on whether a drink is suitable or not. By recognizing the potential impact the labeling of a product and services could have on the likelihood of an under-18 wanting to buy alcohol, the industry was thus recognizing the role it could play in preventing an activity which contravened the law. Complaints of breaches of the Code are adjudicated by an independent complaints panel appointed by the Portman Group.

This mechanism is effective in addressing inappropriate labeling and, to a certain extent the marketing, of alcoholic drinks. It is a relatively simple procedure. The involvement of a third party—the retailer (the off-license, supermarket or pub)—backs up the efforts of the Portman Group because a retailer can refuse to stock a product, which has been declared by the complaints panel as not complying with the Code. This provides an important additional tool for enforcing decisions as long as it is standard practice within the retail sector. The threat of the licensing authority withdrawing the license to sell alcohol from a retailer refusing to follow the Panel's decision also acts as a further motivation, although it has never been implemented.

How successful has the Code been in achieving its stated objective of "preventing irresponsible consumption by people under the age of 18" as well as "combat[ing] all forms of alcohol misuse among all age groups" (or rather, of removing the contribution made to irresponsible consumption by the products and services of the companies involved)? It is clear that the number of complaints made against products and services declined sharply after the first two years of the Code's operation (April 1996–April 1998). In fact, 94 out of the 120 complaints made to date (78%) occurred in this initial period, since when the number of complaints has been in single figures or at zero in all quarters.

Analysis of the grounds for the complaint shows that the most frequently stated reason is that the product and services appeals overtly to under-18s followed by the claims that the alcohol is unclear or that sexual prowess is implied in the product's label (both between 10 and 15 cases). In terms of overall compliance levels, the number of products withdrawn or modified far outweighed those where there was non-compliance. The initial peak appears to indicate that producers, after initial problems with compliance, adjusted their behavior to take the Code's guidelines better into account to avoid recurrent problems with their products and services. That the producers became more interested in how to comply with the Code is illustrated by the high number of requests made to the Group for "pre-launch" advice (these now exceed the number of complaints made).

The Code has been relatively successful in forcing producers to take into account the impact of their merchandising of alcohol on the under-18 age group. The content of the Code is currently under review in order to ascertain whether or not it might incorporate new areas in the future—such as better regulation of happy hours and point of sale promotions—a revised Code is likely during 2002. Actors in other European countries with an interest in this question have also used the Code. For example the recent Recommendations on Drinking of Alcohol by Children and Adolescents adopted by the EU heads of government with the aim of establishing standards in this area drew substantially on the text of the Code and the procedures it follows.

Lessons

The impetus to take action came from the press and public opinion in 1995/6. The action was taken by the drinks industry on its own.[23]

The immediate motivation of the drinks industry was to respond to the criticisms that were appearing in the press and elsewhere. That motivation is likely to have been quite complex, and to have drawn on many of the reasons outlined in our section on corporate responsibility. Many of the heads of the companies are more or less public figures and no doubt do not like to be pilloried. Although it was not then clear whether the government would adopt tighter legislation on the drinks industry (as is now planned), or what that legislation might be, they did not choose to be on the wrong side of the government. The executives of the companies, like the peers, would have been critical of hooliganism and underage drinking. It was their products that were involved, and nobody feels good about making profits by doing wrong. All this adds up to enlightened self-interest.

The law was unimportant, except as a distant threat. But individual premises are subject to the licensing laws, which could have led to the withdrawal of licenses (in practice a rare event) if abuse were flagrant. But the self-regulatory code was at the heart of the solution. All the money for this self-regulatory initiative came from the companies themselves.

Deficiencies

The code of conduct has worked well to regulate the product development and marketing to under-18s by the drink manufacturers. It is harder to deal with cases of hooliganism by drunken adult males who have bought properly marketed and easily available standard drinks. Some such cases have little to do with the promotion of drink, let alone its improper promotion. Innocent availability is enough. It is hard to enforce legislation such as that designed to penalize retailers who continue to sell to those who have already had too much to drink. The manufacturers can do little to prevent the excessive or inappropriate drinking that results from encouragement by pubs and clubs, short of refusal to supply or informing on them to the licensing authority.

This is not to criticize the Portman Group or the measures that it has taken. It is only to point out the limits of such a scheme, however well intentioned and organized, and to note that other measures are needed, over and above such self-regulation, to deal with other aspects of alcohol related crime (Home Office, 2000b).

There is a serious problem about the figures for alcohol-related crime. It appears that the collection of data is incomplete and haphazard (The Portman Group, 2002). It is hard therefore to know what contribution alcohol may make to crime, violence, and injury, and hence what resources and remedies are needed to deal with what may be a quite ill-defined problem.

3. Credit Cards

The Problem

In the early 1990s, there were high and increasing losses from fraud involving lost and stolen credit cards, and lower levels on "mail not received" where cards were intercepted in the postal system. After significant success tackling these aspects of fraud, the mid-1990s saw an increase in the growth of problems with skimming card data and the manufacturing of counterfeit cards.

Business Solution

Improved security and technology through the creation of a secure card authentication method combined with a secure customer verification method.

The Story

Table 1 below shows the substantial losses at the beginning of the decade, falling in the middle 1990s, and rising again in the later years.

Table 1 Association for Payment Clearing Services (APACS) Fraud Losses (£ Millions)[24]

	Other	Card not present	Application fraud	Counterfeit	Mail non receipt	Lost and stolen	Total
1991	1.6	0.4	2.0	4.6	32.9	124.1	165.6
1992	1.0	1.3	1.4	8.4	29.6	123.2	165.0
1993	0.8	1.6	0.9	9.9	18.2	98.5	129.9
1994	0.5	2.5	0.7	9.6	12.6	71.1	96.9
1995	0.3	4.6	1.5	7.7	9.1	60.1	83.3
1996	0.5	6.5	6.7	13.3	10.0	60.0	97.1
1997	1.2	12.5	11.9	20.3	12.5	66.2	122.0
1998	2.3	13.6	14.5	26.8	12.0	65.8	135.0
1999	3.0	29.3	11.4	50.3	14.6	79.7	188.3
2000	6.5	56.8	10.2	102.8	17.3	98.9	292.5

In the early 1990s, a number of steps were taken by card issuers, with the support of retailers, to tackle the problem of lost and stolen fraud and mail not received fraud. Lost and stolen fraud was greatly reduced by the introduction of increased levels of authorization at the point of sale. The fall in mail non-receipt fraud from £33 million in 1991 to just £9 million in 1995 can be explained by the work that the card issuers did with the Post Office at the time to tighten up security regarding the distribution of cards (and PIN numbers) via the postal system.

However, towards the end of the decade counterfeiting together with "card not present" fraud had emerged as the two key new threats. Counterfeiting occurs where a counterfeit card is manufactured using the details on genuine cards. There are several ways in which criminals can get the details necessary to produce a duplicate card: by collecting discarded customer receipts from ATMs; accessing a site on the Internet which generates usable card numbers (neither of which is illegal); or placing an accomplice at a petrol station or restaurant to "skim" card numbers onto a laptop computer out of sight of the customer. Criminal gangs can produce large numbers of fake cards using encoding and embossing equipment (freely available on the open market) and could expect a hit of around £1,500 for each card used.

Card-not-present fraud occurs where genuine card details are obtained and used to perform mail or telephone order fraud from discarded receipts or stolen till rolls. In 2000, £56.8 million was lost by the industry through card-not-present fraud (compared to £0.4 million in 1991) and over £100 million through counterfeiting (£4.6 million in 1991). Barclaycard's own figures suggest that the annual figure for counterfeit fraud has now risen to over £100 million. These two fraud types now account for 55% of all card fraud.

What solutions are the banks developing to meet these new threats? These relate to both the technological and educational aspects of fraud prevention and include:

- "Neural networks" (such as Falcon): these are computer systems that monitor patterns of spending by customers in order to spot any significant deviations that could be a sign of a fraudulent transaction. They are now in widespread use by the banks.
- The introduction of stricter security checks when "remote" purchases are made. This includes new digits on the cards and an address verification service. These measures are designed specifically to combat card not present fraud.
- Introduction of chips in cards and PIN equipment at the point-of-sale.
- Better informing of retailers of the problem of counterfeit fraud and guidance on how to identify fake cards.

The Benefits of Chip and PIN Technology

New cards with chips built in are currently being introduced into the market at a rate of 1–2 million per month.[25] It is estimated that more than 17 million cards with chips are now in circulation. These are harder to counterfeit than the older cards with magnetic strips. The use of the chip also requires the installation of chip-reading terminals at the point of sale in order that the card may be authorised.[26] The card issuers announced in February 2002 the extension to all cards by 2005 of chip and PIN technology at the point of sale.[27] This will help reduce lost and stolen fraud which, after the dip in the mid-90s, is now on the increase again and in 1999 was still the largest fraud category at £80 million.

Barclaycard believes that these measures have the potential to cut substantially lost and stolen and counterfeit fraud.[28] It points to the example of France, where overall levels of card fraud are much lower than in the U.K., because the banking and retail sectors there have been using chip and PIN for almost 10 years. This change was first suggested by Levi in 1991 (Levi & Pithouse, 1991). The difficulties and delay in practical implementation had several causes. The changeover to chip equipment has been fastest in bank terminals because the issuers are usually banks and so can mandate this (over half of the U.K.'s bank terminals are now "chipped").[29] However, there is no easy way of persuading retailers to switch over to chip and PIN technology at the point of sale (and a chip is useless without it). Their reluctance to change has been due to concerns over the cost of replacing their current systems (whose shelf-life may still have several years to run) simply in order to accommodate the new equipment. Another problem has been the belief that the chip and PIN system will slow down transaction times for customers.[30] Above all, the structure of the industry—many card issuers, many retailers, many acquirers—makes agreement hard. Companies that normally compete find it difficult to cooperate. And sheer numbers make it hard to arrive at common ground among different interests. Competition legislation is a complication.

What could be done to encourage the retail sector to come on board? On the one hand, a change in the contractual relationship between the merchant acquirer (the part of the bank which sells the service) and the retailer is due to come into effect in 2005 which will shift the liability for fraud away from the card issuers to the non-chip party. This means that if a retailer has not installed chip and PIN equipment by that date, then the retailer will be liable to pay for the costs of a fraudulent transaction. This rather hefty stick is designed to spur the retail sector into action well before the 2005 deadline. In addition, card issuers established incentive funds for certain retailers to help them pay for the new terminals (which usually cost between £300 and £400 each).[31]

Cross-Border Fraud

A large proportion of U.K. card fraud now takes place across borders (around 35%) and particularly that of counterfeit transactions (60%).[32] So it makes sense to search for regional and international responses to the problem. The European Commission (2001) published a Fraud Prevention Action Plan in February 2001, which looked at ways of achieving progress on this issue within the EU (card fraud is currently costing the EU 600 million Euro a year and grew by approximately 50% in 2000).[33] It argues that a secure system for non-cash payment is essential for the functioning of the EU economy and that partnership and cooperation between relevant public authorities and private parties are crucial if the introduction of new technologies aimed at improving security is to succeed.[34] It concludes that:

> Fraud prevention can only be effective through a combination of coordinated preventative measures and a comprehensive regulatory environment, including adequate sanctions.[35]

Lessons

The impetus for action, the action itself, and the motivation to act derive primarily from the commercial self-interest of the companies themselves. Because in practice the cost of fraud falls on the companies not the customer, even if the customer is at fault, they have strong reasons to reduce fraud in order to protect their profits. But pressure from the government to make the changes to chip and PIN was also key.

Central government and the police have reasons to be motivated to reduce credit card crime beyond the consumers' responsibility to reduce crime generally. This is because credit card fraud is a good way of getting cash, as the skimming example shows, and cash is the lubricant of drug and other organized crime. Gangs often have links to organized criminal groups involved in other activities such as drugs smuggling, bootlegging and money laundering.

The law, in the sense of legislation directed specifically at requiring companies themselves to address these types of crime, has been unimportant. All of the money to develop and implement the technological solutions has come from the companies and retailers. The partnership required is unusual but central. The whole industry—card issuers, acquirers, retailers—has to agree on changing technology at around the same time to the new, PIN and chip technology. This coordinated change is not easily achieved through the market by arms length contractual negotiations.

Deficiencies

The response by the police to credit card crime has depended a great deal on the force in question. The 1991 APACS recommendation that all police forces have a dedicated check and credit card fraud squad has generally not been put into effect (Levi & Handley, 2001, p. 21). There are, according to the industry, circumstances when the issuing company finds itself doing all the work to prepare a case for prosecution, a task for which the police and CPS would normally take the lead.

At present there is not under English law (as there is in the U.S.A.) an offence of identity theft. Many credit card frauds involve the thief assuming the identity of the legitimate cardholder. This issue is a general one for electronic crime.

The possession of card skimming and encoding equipment is strong evidence of criminal intent, since there is no legitimate use to which they can be put. Yet it is not a crime to own, sell or manufacture such items.

5. CONCLUSIONS AND RECOMMENDATIONS

In each case study, we have set out the issues that we think most relevant to the initiative's success. It is apparent that each case differs importantly from the others, and that there is no one-size-fits-all template or rule book which will tell us how to set about achieving similar success in other industries. But our task now is to draw out the policy conclusions about how to achieve more in this area and tap the full potential of companies' contribution to crime reduction. These are based for the most part on our case studies, but also on other examples and evidence from our research.

It is clear that there are many successes, and they are substantial. We are not talking about a minor contribution here and there, welcome but not decisive. Car crime is down because of immobilizers. Credit card fraud sank when interception in the post was stopped. Our research confirms that what companies do in this field matters a great deal.

Companies are central in another way. When something has to be done, and when their participation is secured, they are necessarily in the front line of prevention, because they produce the goods and services which have to be modified. It is they who can best work out what to do, and they who can do it. The other players matter, but in this area the companies are primary.

There is very little recognition of this central role by Government, the press, the police or the public. Public awareness could be significant if it influenced the customers' choice of products or services, for instance which mobile phone operator they used. But "theft-ability" is not one of the criteria companies use in promoting their products. It is the companies who meet most of the directly identifiable costs of these crime reduction measures. Who pays to redesign a product, or increase security, can be the key issue, as it was in the Eurotunnel case.

Thus far, whether action is taken has depended in the first instance on whether the crime is salient in the minds of the public, the press, and hence Ministers. If lager louts or hooligans are a scandal, then it is likely that action will be taken. In that sense, action in the past has been reactive, particular and opportunistic. It has not been based on any mainstream notion that goods and services should stop facilitating crime. Accordingly, action on the redesign of goods and services has generally, because reactive, been backward looking. Retrofitting is the natural remedy, because the initial design of the product or service was determined some years ago before the consequences for crime were salient.

The solution is slow to arrive. The cycle of the problem occurring, being identified, getting the salience needed to require a solution, finding the solution, putting it into effect, can be painstakingly prolonged. Partly because the problem is everybody's and therefore nobody's, time is needed for responsibilities to be recognized, for cooperation between organizations to affect change and to secure commitment from others not to free ride.

Not much time has been spent on trying to arrive at general principles of corporate social responsibility in this area. A problem arises, it demands solution, and the companies take action because there is pressure on them to act, and because they think that they ought to. There is evidence, not only from the case studies but also from other examples such as the attempted theft of service in the telecommunications industry, and drive-away theft of petrol, that crimes against business are low priority for investigation and prosecution as compared with crimes against other citizens. The latest British Chambers' of Commerce study (2001) confirmed this.

Companies, as the BCC survey showed, are quite reluctant to report crime. This is in part because they do not think that anything will happen if they do. It may also be that crimes such as fraud reflect badly on their competence and reputation. For this and other reasons, statistics on business crime are weak, and the assessment of such initiatives as Retail Crime Partnerships is made harder. More generally, there is a pervasive problem of data and monitoring of crime related to products and services. The examples of alcohol and mobile phones show how public alarm allied to doubtful statistics can in these cases, as with crime generally, lead to confused understanding of what the problem is and how it should be tackled.

Of course, the law matters as the procedure whereby crimes are defined, prosecutions take place, and offenders sanctioned. But legislation to require business cooperation in addressing crime is rare. An obvious case is the EU legislation on car immobilizers. But that served more to ratify a process that the companies had already started than to start the process off. There may be an assumption that regulatory requirements are inappropriate, or more likely that little thought has as yet been given to the need for business cooperation and how to secure it.

The role of central government is significant. Sometimes in an obvious way: knocking company heads together to cajole them to act. Such arm-twisting may not be constitutionally attractive—it is a way of imposing quasi obligations on companies without going through any proper process—but it has been effective. The 1998 Crime and Disorder Act, requiring local partnerships, has provided a significant context for business involvement. Perhaps more important is the somewhat low key provision within Whitehall of networks and forums where relevant parties—the industry, officials, academics—can work out together what to do without any elaborate or preconceived formulae. Initials such as VCRAT and PCRAT may not make the journalist's blood race. But these bodies have been central to success.

Solutions are not permanent. The criminal and the technology move on. Nobody could have anticipated the possibility of the theft of credit card details by skimming. The motivation and the action to prevent the crime must continue after the first fix.

Partnership is a real and important condition of success. All the cases require co-operation between players with different skills and different interests, with no clear rules about how they are to work together. These successes could not have been achieved solely by arms length contractual arrangements across a market. They require working out as you go along what would be best for which player to do. These problems are poorly adapted to the clear specification of roles *ex ante*. See, for example, the unintended but admirable incorporation of the non-statutory Retail Crime Reduction Partnerships into the framework of the 1998 Crime and Disorder Act.

For the partnership to work everyone has to have a reason to take part. This is not a trivial condition. It is not enough to wish in a general way that crime were lower, or that someone else would do something about it. Nor is it enough to assert that one party or the other, particularly companies, are responsible on the basis of some general principle. Each player has to want and to be able to do something about it. Reasons, or incentives, are central to all partnership solutions to crime. Without them there will be no action, and the strategy degenerates into a wish list.

There has been little awareness or work among consumer groups of the role that companies can take in designing out crime, and the interest that customers have in buying secure products. This is a missed opportunity.

Finally, speed. The examples of credit cards and mobile phones emphasize that it can take a long time, in the case of credit cards a very long time, to get to the solution. The recognition lag, that there is a problem; the solution lag, what should be done about it?; the responsibility lag, who should take action?; the implementation lag; all these add up to years of delay, while the crimes continue. So even when as in these cases the final result is good, we have to ask whether it could not have been arrived at quicker. It is always easy for the irresponsible commentator, using hindsight, to complain about such delays. And certainly no system is going to produce instant solutions to often novel problems. But we hope that our recommendations may help to speed up the process, by providing policy principles and machinery which make it more likely that problems will be identified and acted on in advance, rather than waiting until salience the press and publicity force action.

Way Forward

Our recommendations centre on the need to generalize the principles that lie behind the past successes of company involvement in crime prevention. It would be good if government and industry were to continue to work together, opportunistically, when particular cases of "hot goods" or services are identified and need fixing. It would be better if action were to take place earlier, before the problem arises, by each company planning to reduce the opportunities for crime which new goods and services present.

It is not possible to say now what products and services will cause problems in the future. Government and companies therefore need mechanisms which plan for what may happen, as well as continuing to deal with the present. If that is not accepted, we are left with the piecemeal *ad hoc* solution of high profile problems, such as joyriding or mobile phone theft, without tapping the full-scale potential of what business can do. That would be a missed opportunity.

The Government has to decide, as a matter of policy, whether the analysis presented in this paper—that business has a central role to play in crime prevention, going beyond the piecemeal contributions of individual industries—is right. If there is not agreement on this central point, then there is no purpose in pursuing the matter further. We do not believe that further evidence is needed to resolve this.

Equally important, it must be recognized that crimes against business should be taken seriously. It is the responsibility of the state to deal with them when they occur, and of businesses to take seriously their role in preventing and reporting them. They are not to be treated as just another cost of doing business. Nor are they victimless crimes. The businesses that are the victims are often small proprietors who suffer just as individuals do when they are damaged by crime. Employees suffer from assault and hooliganism and threats whether they work for a corner shop or for a high street chain. Big companies lose money from higher premiums and other costs, and customers too suffer from higher costs through higher prices.

These should be the principles of the partnership between business and the state on which future action will be founded. They are based on the idea that this partnership, like all partnerships, depends not only on the ascription of specific responsibilities, but on a deal from which each benefits and to which each contributes.

Principles of Partnership

Where it is possible to make a material difference to the incidence of crime relating to their goods or services, business will make appropriate changes to their specification, either at the design stage, or by subsequent modification.

- Such modifications must be reasonably practicable, technically and financially. They must not require excessive cost or any material reduction in customer appeal.

- The state and its agencies recognize that such crime prevention is a joint enterprise, in which it too has responsibilities, matching those of business.

- The state agrees that it will fulfill its obligations, through the police, criminal justice agencies and other relevant public bodies, to deal with crimes against business as effectively as any other crimes of similar seriousness.

- Businesses will, as good citizens, routinely report business crime to the police.

All partnerships involve a more or less explicit deal, with each side giving here and taking there. It is that trade in mutual advantage that provides the incentive to make this partnership work. Business recognizes a new responsibility for helping to reduce crimes which do not necessarily affect them but which arise from their products or services; and the state recognizes anew its old responsibility for doing what it should about crimes against business.

Companies Against Crime Strategy

This is the background against which particular initiatives need to be considered. Our recommendations draw on the framework set out in our section on corporate responsibility, where we clarify the variety of reasons that companies have to contribute to social goals such as crime reduction. Unsurprisingly, given our emphasis on partnership and voluntarism, the participants in these proposed initiatives vary from central government to companies to trade associations to professional bodies. A major role of the Companies against Crime Champion, which we advocate below, would be to promote initiatives of this kind, and to generate others.

- Companies should consider the potential for all their new products and services to be used for, or contribute to, criminal offences and should mainstream crime prevention into their design. Where existing products and services are contributing to crime, companies should consider the feasibility of modifying the product or service to reduce that potential. In line with the principles set out above, such modifications must be reasonably practicable, technically and financially. They must not require excessive cost or any material reduction in customer appeal.
- The Government should make a public commitment that the participation of business in crime prevention is a major plank in its crime prevention strategy and, in cooperation with small, medium and large companies, draw up a partnership strategy, Companies Against Crime, to develop their contribution.
- At the same time, the Government should review the priority given in national policy to crimes against business by the police and consult business on areas of concern that they have. It is an important test of the proposed policy that the state should treat crimes against business as they treat other crimes.

- The strategy should primarily be based on the principles of partnership and voluntarism. It should rely on guidance and incentives, with regulation only as a last resort. In the exceptional cases where regulation is needed to require companies to co-operate, whenever possible it should be targeted at those companies that are not meeting their responsibility to address crime rather than the whole sector (for instance at those pubs and clubs that encourage or fail to deter the excessive drinking that spills over into violence on the streets, not the responsible licensed premises in the same area).

- The Companies Against Crime strategy will require mechanisms and institutions at the national and local level for managing the change from opportunism and retroaction to a comprehensive approach (including a unit in the Home Office with responsibility for developing the strategy), working closely with those promoting Corporate Social Responsibility in the DTI. There must be strong and effective national and local forums in which interested parties, from business to government and researchers, can meet on neutral ground to identify future crime developments and devise timely, accepted and practical responses. At the local level, the existing Crime and Disorder Reduction Partnerships could fulfill that role well.

- The strategy requires a conspicuous Companies Against Crime Champion, such as that which was available in the Crime Prevention Panel (CPP) of the Foresight Program, a panel located within Whitehall, bringing together representatives from government and the private sector, chaired by a business leader.

- The first task of the Champion should be to identify the range of incentives and soft regulation that will encourage companies across the full range of manufacturing and service providers to mainstream crime prevention into their product and service design. Secondly, it should identify innovative ways in which companies could contribute to the reduction of specific crimes of particular public concern. Proposals such as those listed in Cracking Crime Through Design should continue to be high on the agenda (Pease, 2001).

- The existing Order under Section 5 of the Crime and Disorder Act should be amended to make specific and clear reference to business as among the parties which local authorities and the police should consult.

This will ratify the substantial part which business does and should play in the identification of crime reduction opportunities, through the statutory Crime Audit and Strategy and through Crime Reduction Partnerships in the execution of such strategies. It will also emphasize the increased role which business could play in the next round of local crime audits.

- In order to ensure greater awareness of the successful role of business in crime reduction, successes such as those in our case studies should appear in companies' Annual Report and Accounts. This should be achieved by disclosure in the statutory accounts of the crime reduction contribution of the company in the preceding year. Such reports, which would be in line with the new emphasis on reporting in the Company Law Review, would detail failures as well as successes. This follows the lead given by the Turnbull report, which recommends how companies should deal with risks, *inter alia*, to their reputation. It would make failure to consider crime prevention a reputational risk.
- Similarly, the Association of British Insurers should consider including failure to prevent crime among the social, environmental and ethical matters on which companies should make disclosure.
- Business in the Community has at present four major areas of work: workplace, marketplace, community and environment within which it runs specific campaigns. Recent activities have focused on homelessness and community and rural regeneration. It should consider the launch of a *Business Action on Crime* campaign to include all elements of the deal which we outline above—pursuit of crime against business and reduction by business of crime.
- There should be greater public recognition of the role of business in this area. For example, contribution to crime reduction should be recognized by the government as a mainstream criterion for the award of honors, alongside the familiar categories of politics, arts, sport and contribution to the community. Similarly, the Queen's Award to Industry categories should be extended to include contributions to crime prevention. The British Security Industry Association should consider providing annual prizes to companies for crime prevention.
- Information on crimes against business requires much improvement. Business must be required to produce figures on the crime of which

it is the victim. The criminal justice system cannot do its job if it does not know how much fraud or theft there is. Crimes against business should appear as a separate category in the British Crime Survey.

- It is for the Home Office to take the lead in improving the data, evaluation and monitoring of initiatives in this area, to deal with problems which our study has identified (Pease, 2001, p. 53). The model put forward in the Marsh report for the Portman Group (2002, p. 43) may be a model of what ought to be done, not only for alcohol related crime.
- Following the example of the Secured Car Parks and Safer Shopping Awards schemes, there should be introduced an ISO standard for security of product and services analogous to the ISO for environmental protection. This would include a kite-marking scheme to enable customers to choose products and services that are "crime proof"—designed to reduce the opportunities for crime.
- Businesses should be encouraged to identify ways of cutting crime that do not involve resort to the criminal law, such as the use of the civil law by retailers against shoplifters,[36] which avoids the harmful consequences to society of ever greater numbers of people with criminal convictions (already standing at 30% of the male population).
- The Government could, in turn, consider ways to assist companies in reducing the potential for their products to lead to crime. For instance, there is no good reason why a member of the public should be in possession of card skimming and encoding equipment. The same arguments apply to the possession of car immobilizer and radio decoding equipment. In a technologically rich age, criminals need a variety of tools of a specific character to counter security technology. It may be possible to identify other kinds of equipment that should not be legally available.
- Finally, a recommendation to the public. Nothing would do more to promote our agenda than the emergence of a public pressure group analogous to those in the environmental field, campaigning for action on products and services of the kind we have proposed. That is how the environment agenda became central to government, media and business thinking. Such a consumer group could identify the ways in which products and services play a part in criminal behavior, advise companies of the ways in which it might reduce that potential, highlight

those that failed to take appropriate action, take test cases where the criminogenic properties of a product have resulted in harm to an individual, and advise government on the public policy initiatives it could take to encourage companies to contribute.

What companies have done to reduce crime in Britain over the last decade has been scarcely recognized. Their record of success, in partnership with others, is striking. Unusually in the criminal justice system, the successes have been cheap for the exchequer (if not for business). What is needed now is a policy framework, the mechanisms, and the will, to do much, much more of the same. This policy area offers a large and untapped opportunity to reduce crime. Our recommendations are designed to help make that happen.

Editors' note: This chapter is a slightly edited version of a report originally published by the Institute for Public Policy Research (IPPR) as one product of the work of the Criminal Justice Forum 2002. We are most grateful to IPPR for allowing us to include the paper in this volume.

Address for correspondence: Institute for Public Policy Research, 30–32 Southampton Street, London WC2E 7RA, United Kingdom; tel +44 (0)20 7470 6100; fax +44 (0)20 7470 6111; email info@ippr.org.uk; www.ippr.org.uk

Acknowledgments: The authors would like to thank the following people who generously gave their time to contribute ideas and information for this report: Ken Pease and Gloria Laycock at the Jill Dando Institute of Crime Science, Martin Warwick of Barclaycard, Mike Schuck at the BRC, James Abraham and the Security Team at Ford Motor Company, Richard Davis and Roger Bourne at the DTI, Penella Price at the Home Office, Jean Coussins of the Portman Group, Bridget Hutter at the LSE, Alison Huxley at the Design Council, Tony Routledge of One-2-One, Tony Burden of ACPO, Lord Sharman of the Foresight Crime Prevention Forum, Olivia Lankester of Friends Ivory and Sime, and Malcolm John and Mary Harris at the Lattice Foundation. Without their substantial help this paper could never have been published. We would like to thank the Design Council for their support of this specific piece of work. We would also like to thank the numerous other people who we spoke to from the retail, credit card, mobile phone and car industries, as well as from the Home Office, DTI, and the police, all of whom aided us greatly. Finally, our thanks go to IPPR's own Criminal Justice Forum for their support and comments on earlier drafts of the paper. IPPR would like to thank Barclaycard, The Bar Council, the Calouste Gulbenkian Foundation, the Esmee Fairbairn Foundation, Group 4 Falck and the Institute of Legal Executives support of the Criminal Justice Forum's work. It must be pointed out ...st we have benefited from much advice and assistance from many quarters, the views expressed in this report are the responsibility of the authors.

NOTES

[1]See: www.britishchambers.org

[2]"Similar binding" is needed to include rules such as the Stock Exchange Code in Britain (which are not statutory but nevertheless very powerful), codes of conduct, and similar devices.

[3]Which can also be seen in terms of conventional economic theory as the territory where there are externalities to business behavior that are not covered by the law.

[4]The speech of DTI minister Douglas Alexander to the Association of British Insurers (see: www.abi.org.uk) is an example. His emphasis is on the clear business benefits of responsible behavior, and of the possibilities of modifying behavior by revisions of company law and pension regulations.

[5]There is certainly a collectivist version of this argument that works. But that is not the way it is usually presented.

[6]International Transport Roth GmbH & Others and The Home Office. Many of the same issues arise in relation to the disagreement between the British Government and Eurotunnel over security, particularly concerning the freight terminal near the Sangatte asylum seeker centre: what the ferry companies should do to stop similar stowaways: and the rules which the airlines have to follow to prevent them carrying passengers without documentation, or drug smugglers.

[7]Ibid., para 5.

[8]For the formidable provisions of the Code, see ibid., para 18. For the judge's views on its reasonableness, see paras.139 and 140.

[9]Ibid., para 156 ff.

[10]Ibid., paras 43 and 44.

[11]Ibid., para 62ff.

[12]Ibid., para 41.

[13]But this is a very cursory account indeed of the problems compared with the 131 pages of the judgment.

[14]Harrington and Mayhew, 2002; *Observer* January 6, 2002; *Guardian* January 9, 2002.

[15]See: www.britishchambers.org.uk

[16]See also: www.crimecheck.co.uk for details of a number of similar surveys of the level of business crime.

[17]Similarly, visible VINs need to be built in to the windscreen in such a way that prevents their being obscured by objects placed on the dashboard.

[18]It costs around £75 to fit an approved electronic immobilizer.

[19]On February 11, 2001 John Denham, the Minister for Crime Reduction, announced that the scheme now covers 1,000 car parks and that government support in 2002/2003 was to be increased to £300,000 (sic).

[20]The BCS shows a fall of 11% in vehicle related crime between 1999–2001 (Home Office, 2001b).

[21]The Motor Insurance Repair Research Centre.

[22]The Act provides for: Regulation of Motor Salvage Operators—introduces powers to regulate the motor salvage industry and require motor salvage operators to register with local authorities, keep records and for the police to have right of entry to registered premises without warrant; and Regulation of Registration Plate Suppliers—introduces powers to control the supply and issue of number plates. It requires number plate suppliers to register, to make suitable checks before selling a number plate and to keep records of transactions; and Other Provisions Relating to Vehicle Crime: such as it enables a vehicle which has been written off by an insurance company to be required to have an identity check if the Driver and Vehicle License Agency (DVLA) receives a request for it to be allowed back on the road. This will prevent the identity of stolen vehicles being disguised by that of other, legitimate, vehicles. It also provides a power to prescribe the form and manner in which information on number plates must appear.

[23]Action to do with the self-regulation of sensible drinking. Of course, others have since done a great deal: see the latest strategy on alcohol and violence. But this particular initiative seems to have been undertaken by the companies alone.

[24]www.apacs.org.uk

[25]Barclaycard estimate.

[26]At present most cards with chips also have a magnetic strip allowing either authorization method to be used depending on the terminal.

[27]"PINS to replace signatures," APACS press release, February 8, 2002.

[28]Interview with Martin Warwick, Barclaycard.

[29]Barclaycard's figures.

[30]The British Retail Consortium says that "evidence from the USA shows that PIN at the point of sale is quicker than signature verification by about two seconds." See Home Office Press release February 8, 2002.

[31]Interview with Barclaycard.

[32]Ibid.

[33]European Commission Communication, 9/2/01, "Preventing fraud and counterfeiting of non-cash means of payment," p. 2.

[34]Ibid, p. 3.

[35]Ibid, pp. 3–4.

[36]See "An Academic Interest" in *Communicate*, June 2001, pp. 18ff.

REFERENCES

APACS. (2001). *Fraud in focus: An update on measures to prevent plastic card fraud.* London: APACS (Administration) Ltd.

APACS. (2000). *Card fraud: The facts.* London: APACS (Administration) Ltd.

Association of British Insurers. (2001, October). *Disclosure guidelines on socially responsible investment.* Retrieved from www.abi.org.uk

Ballintyne, S., Pease, K., & McClaren, V. (2000). *Secure foundations: Key issues in crime prevention, crime reduction and community safety*. London: Institute for Public Policy Research.

(The) British Chambers of Commerce. (2001, December). *Business Crime Survey Executive Summary*. Retrieved from www.britishchambers.org.uk

British Retail Consortium. (2001). *8th retail crime survey 2000*. London: The Stationery Office.

British Retail Consortium. (2000). *Retail crime survey 1999*. London: The Stationery Office.

Burrows, J. (1988). *Retail crime: Prevention through crime analysis*. Crime Prevention Unit: Paper 11. London: Home Office.

Communicate. (2001, June). Issue 11, pp. 7–11, 18–22, 32–33.

Communicate. (2001, Autumn). No 12.

Communicate Scotland. (June). Issue 11, pp. 10–14, 19–24, 29.

Craig-Smith, N. (1990). *Morality and the market: Consumer pressure for corporate accountability*. London: Routledge.

Crime and Disorder Act 1998. Retrieved from http://www.legislation.hmso.gov.uk/acts/acts1998/98037-b.htm

Department of Trade and Industry (DTI). (2001). *Business and society: Developing corporate social responsibility in the UK*. Retrieved from http://www.societyandbusiness.gov.uk

DTI. (2000). Foresight Crime Prevention Panel. *Turning the corner*. London: DTI.

Ekblom, P. (2000). *Less crime by design*. Lecture to Royal Society of Arts. Retrieved from www.edoca.net/Resources/Lectures/Lesspercent20Crimepercent20by/20Design.htm

European Commission. (2001). "Preventing Fraud and Counterfeiting of Non-cash Means of Payment," COM (2001)11. Brussels: European Commission.

Farrington, D. P., Gallagher, B., Morley, L., St. Ledger, R. J., & West, D. (1986). Unemployment, school leaving and crime. *British Journal of Criminology, 26*(4), 335–356.

Faulkner, D. (2001). *Crime, state and citizen*. (See especially chapter 18, "Community Safety: Preventing Crime and Disorder.") Winchester: Waterside Press.

Foresight Crime Prevention Panel. (2000). *Turning the corner*. London: DTI.

Foresight Crime Reduction Panel. (2001). *Criminal justice: The way ahead*. Cmnd. February Annex G.

Fox, C. (2000). Social responsibility and crime reduction. Published in the annex of Foresight's *Turning the corner*.

Friends, Ivory and Sime [Co.]. (2001). *Corporate social responsibility: Report for the year ended 31 December 2001*. London: Author.

Garland, D. (2001). *The culture of control: Crime and social order in contemporary society*. (See especially chapter 6, "Crime Complex: The Culture of High Crime Societies.") Oxford: Oxford University Press.

Gunningham, N., & Grabosky, P., with Sinclair, D. (1998). *Smart regulation: Designing environmental policy*. Oxford: Clarendon Press. (See especially the chapter on "Instruments for Environmental Protection.")

Harrington, V., & Mayhew, P. (2002). *Mobile phone theft*. Home Office Research Study 235. London: Home Office.

Home Office, Retail Crime Reduction Action Team. (1998). *Community Crime reduction partnerships: The retail contribution—partnership documentation* London: Home Office.

Home Office. (1999a). *The government's crime reduction strategy*. London: Home Office. Retrieved from http://www.crimereduction.gov.uk/crssummary.htm

Home Office, Vehicle Crime Reduction Action Team. (1999b). *Tackling vehicle crime: A five year strategy*. London: Home Office Communications Office.

Home Office. (2000a). *Preventing customer fraud—a guide for retailers*. Retrieved from http://www.homeoffice.gov.uk/crimprev/cppcf.htm

Home Office. (2000b). *Tackling alcohol related crime, disorder and nuisance—Action Plan*. Retrieved from http:www.homeoffice.gov.uk/pcrg/aap0700.htm

Home Office. (2000c). *Vehicle crime: Car theft index*. Retrieved from http://www.homeoffice.gov.uk/crimeprev/vc_index.htm

Home Office, Lord Chancellor's Department, Attorney General's Office. (2001). *Criminal justice: The way ahead*, CM 5074. London: The Stationery Office.

Home Office. (2001a). *Policing a new century: A blueprint for reform*. CM 5326 paras 5.44—5.50.

Home Office. (2001b). *The 2001 British Crime Survey*. London: Home Office.

Home Office. (2001c). *Vehicle Crime Reduction Action Team—VCRAT Recommendations: Progress October 2001*. Retrieved from http://www.crimereduction.gov.uk/vcrat6.htm

Home Office. (2001d). *Crime reduction: Secure design*. Retrieved from http://www.crimereduction.gov.uk/securedesign8.htm

Home Office. (2001e). *The government's crime reduction strategy: Reducing burglary and property crime*. Retrieved from http://www.homeoffice.gov.uk/crimprev/crsdoc4.htm

Home Office. (2001f). *Community crime reduction partnerships: The retail contribution*. London: Home Office Communication Directorate.

Home Office (date unknown). *Tackling property crime: An initial report from the Property Crime Reduction Action Team*. Retrieved from http://www.crimereduction.gov.uk

HR Connect Ltd. (2001). *Employing people with conviction*. London: Chartered Institute of Personnel and Development.

Hutter, B. M. (1997). Regulatory relations: Reforming regulation. In W. S. Lofquist, M. A. Cohen, & G. A. Rabe (Eds.), *Debating corporate crime*. London: Anderson Publishing.

ICAEW (The Institute of Chartered Accountants in England & Wales). (1999, September). *Internal control—Guidance for Directors on the Combined Code*. London: The Institute of Chartered Accountants. (See www.icaew.co.uk)

Joseph, E. (2001). *A new agenda for business*. London: IPPR.

Joseph, E., & Parkinson, J. (2001). *Confronting the critics*. London: IPPR.

(The) Judge Institute of Management Studies. (2000). *Design against crime: A report to the Design Council*. The Home Office and the Department of Trade and Industry. Cambridge: Cambridge University Press.

Laycock, G. (2001). *Scientists or politicians—who have the answer to crime?* Jill Dando Institute of Crime Science lecture, delivered 26th April 2001.

Levi, M., & Handley, J. (1998). *The prevention of plastic and cheque fraud revisited.* Home Office Research and Statistics Directorate. London: Home Office.

Levi, M., & Handley, J. (2001). *Criminal justice and the future of payment card fraud.* London: IPPR.

Levi, M., & Pithouse, A. (1991). *The prevention of cheque and credit card fraud.* Crime Prevention Paper 26. London: Home Office.

Office of the Data Protection Commissioner. (2000). *CCTV Code of Practice.* Wilmslow, Cheshire: Office of the Data Protection Commissioner.

NACRO. (1981). *Unemployment and young offenders in Northern Ireland.* London: NACRO.

NACRO. (1999). *Going straight to work.* London: NACRO.

NACRO. (2001). *Drink and disorder: Alcohol, crime and anti-social behavior.* London: NACRO.

Pease, K. (2001). *Cracking crime through design.* Design Council Policy Paper. London: Design Council.

(The) Portman Group. (1998a). *The Portman Group: Promoting sensible drinking.* London: Portman Group.

(The) Portman Group. (1998b). *Keeping the peace: A guide to the prevention of alcohol-related disorder.* London: Belmont Press.

(The) Portman Group. (2000). *The Portman Group Code of Practice on the naming, packaging and merchandising of alcoholic drinks.* London: Portman Group.

(The) Portman Group. (2002). *Counting the cost: The measurement and recording of alcohol-related violence and disorder.* London: The Portman Group.

Scottish Executive Central Research Unit and Scottish Business Crime Centre. (n.d.). *Counting the cost: Crime against business in Scotland.* Edinburgh: Scottish Executive Central Research Unit and Scottish Business Crime Centre.

Schuller, N. (2000). The business case for crime prevention. Published in the annex of Foresight's *Turning the corner.*

The Sunday Times. (2000, November 26). Pin numbers to replace credit card signatures.

Promoting Design against Crime

by

Simon Learmount
Judge Institute of Management, University of Cambridge

Abstract: *This paper reports the findings of research that aims to cast light on the current state of crime-resistant design in the U.K. It identifies the various factors that directly or indirectly influence the* capacity *(for example, competence, resources, training, and education) and* motivation *(for example, incentives, self-interest, legislation, social conscience) of designers and businesses to incorporate crime-resistant features into the design of products, in order to make some initial recommendations as to how this might be improved. First examined are military design and ecodesign, to ascertain whether there are any lessons that might be usefully transferred to the area of crime-resistant design. Considered next is how crime resistance is currently incorporated into design education and practice, and what might be done to raise awareness of the issue amongst design educators and designers themselves. This is followed by an examination of the new product/service development process in various companies operating in six different sectors—car manufacture, train carriage manufacture, school commissioning, house building, e-commerce and consumer electronics—in order to understand how best to persuade industry to take greater account of the crime resistance of their products and services. Next, a survey of consumer attitudes towards crime-resistant design is reported; these are thought to be of key importance in encouraging businesses to design*

Crime Prevention Studies, volume 18, pp. 141–178.

against crime. Finally, the main conclusions are drawn together to make some suggestions as to how crime-resistant design might be facilitated.

1. INTRODUCTION

In recent years the impact that effective design can have in reducing crime has been increasingly recognized. Think about the security features which now appear on the majority of modern cars—improved locks, alarms, immobilizers, tracking systems—which are widely thought to have contributed significantly to the reduction in the U.K.'s car crime over the past 10 years. Certainly, it has been shown that older cars are at significantly greater risk of theft.[1]

Several Home Office reports have suggested that crime-resistant design might influence crime levels in many other areas, and yet products and services continue to be developed with little regard to their potential effect on criminal opportunities and activity. Why is this so—and what can be done to change the situation?

With these questions in mind, the Home Office, the Department of Trade and Industry and the Design Council commissioned researchers at the University of Cambridge, Sheffield Hallam University and the University of Salford to undertake a program of work on *Design against Crime*. The specific brief was to document the current awareness and state of crime-resistant design in industry and the design education world, to identify factors that constrain or facilitate the use of design best practice in counteracting crime, and to suggest how the existing situation might be improved. In particular, the aim was to explore the various factors that directly or indirectly influence the *capacity* (for example competence, resources, training and education) and *motivation* (for example incentives, self-interest, legislation, social conscience) of designers and businesses to incorporate crime-resistant features into the design of products and services. The main focus is on the design of *secure products*—which are inherently resistant to crime—rather than *security products*, such as add-on steering wheel locks for cars. Furthermore, in this context, by products is meant not just consumer goods: for example, buildings and environments are of equal importance to this study.

This paper provides a summary of the findings from this research, which are discussed in more detail in the report, *Design Against Crime*

(Learmount et al., 2000). It begins with a brief review of Situational Crime Prevention (of which crime-resistant design is an integral part) and the various claimed advantages and limitations of that approach. Section 3 briefly explores military design and ecodesign, to see how design has already been used in these fields to address similar issues to those being investigated here, and to ascertain whether there are any lessons that might be usefully transferred to the area of crime-resistant design. Section 4 examines how crime resistance is currently incorporated into design education and practice, and suggests ways of raising awareness of the issue amongst design educators and designers themselves. Section 5 explores the new product/service development process in companies operating in six different sectors (car manufacture, train carriage manufacture, school commissioning, house building, e-commerce and consumer electronics), in order to understand how best to persuade industry to take greater account of the crime resistance of their products and services. Section 6 discusses the awareness and attitudes of consumers towards crime-resistant design, which are argued to be key factors in encouraging businesses to design against crime. Finally, section 7 draws together the main conclusions from the research, and discusses the recommendations that follow from the findings.

In sum, the research strongly supports the idea that crime-resistant design can complement other approaches in reducing crime, and moreover is likely to have a positive, cost-effective impact in a relatively short time-frame. A concerted attempt to improve awareness of "Design Against Crime" in industry and amongst designers, and to encourage design decision makers to take crime-resistant design seriously, would appear to be a valuable and worthwhile initiative, which could ultimately have significant and long-lasting benefits for the U.K. At the moment though, it was found that there are wide variations in the extent to which crime-resistant design is incorporated into design education curricula or companies' product or service development processes. In design education, although exemplars of designing in crime resistance do exist, general awareness of the impact that design can have on criminal activity and opportunity is very limited. In the case of industry, many companies in the automotive and e-commerce sectors are fairly advanced in incorporating crime-resistant design into their development processes, but organizations in other sectors lag way behind. The report identifies the reasons why these variations exist, and makes clear recommendations as to how the situation could be improved.

2. SITUATIONAL CRIME PREVENTION

Since the late 1970s, Situational Crime Prevention (SCP) has been increasingly recognized as a viable strategy for reducing crime. Based on the premise that much crime is contextual and opportunistic, situational initiatives typically alter the context to diminish opportunities for crime. Conceptually, SCP is not new: people have always responded to perceived risks by altering their behavior to reduce the risks. For example doors get locked, windows shuttered, dogs purchased, and alarm systems installed in order to make an intending offender's work more difficult. What is different about SCP as a crime reduction strategy is that it involves a systematic, strategic effort to develop, test and implement techniques to reduce the opportunity for crime.

SCP, if successful, typically has immediate benefits. It does not aim to affect offenders' propensities or motives. It takes these as given, and proceeding from an analysis of the circumstances giving rise to particular crimes, introduces specific changes to influence the offenders' decision or ability to commit these crimes at particular places and times. Thus it seeks to make criminal actions less attractive to offenders, rather than relying on detection, sanctions or reducing criminality through, for example, improvements in society or its institutions. This approach can be applied to any environment, product or service that is a potential target for crime.

The effect of changes to the crime situation can be to *deter* (making crime more risky), *discourage* (making crime harder or less rewarding), and/or *remove excuses* (awakening the conscience, as in the example of "shoplifting is stealing"). SCP techniques typically involve modifying a crime *target* that is vulnerable and attractive, a *crime preventer* who is absent or incapacitated, or a *crime promoter* who is negligent or actively supports a criminal act. SCP can apply to every kind of crime, not just to "opportunistic" or acquisitive property offences: there are examples of successful applications to more calculated or deeply-motivated crimes (e.g., hijacking, homicide and sexual harassment), as well as ones committed by hardened offenders (Clarke, 1997).

There are criticisms of SCP methods though. For example, crime is constantly evolving: recent social and technological change has brought new targets (laptop computers, mobile phones), as well as tools (cordless drills) and means of disseminating criminal techniques (for example the Internet) which help offenders develop countermeasures. It is argued that SCP does have the potential, however, to work both *reactively* to emerging crime problems as well as in *anticipation*, through crime impact analyses on proposed new policies, practices and products, and through incorporating prevention within the design process (Ekblom, 1997).

One of the most pervasive criticisms of SCP is that it tackles the symptoms of crime rather than its causes (Laycock & Tilley, 1995). It is also argued that by focusing on products, the human dimension of crime is ignored—it is people who own products that are stolen or damaged, who buy stolen goods, who are defrauded or attacked, and who use or fail to use security features. From this perspective, the reduction of criminal opportunities is seen as a blocking mechanism—a mechanistic way of preventing crime from happening while not affecting the motivation to offend. Many criminologists, nevertheless, suggest that opportunities do to some extent cause crime: the decision to offend or not can hinge on relatively minor events that push the individual in one direction or another. For example, the combination of alcohol, a group of excitable peers, and a vulnerable vehicle could well increase the risk that an otherwise law-abiding young person commits car crime.

There is also some debate whether situational approaches *prevent* crimes or merely *displace* them to other, less-well protected targets, times and places. Recent reviews suggest that although this may happen to some extent, displacement is limited (Hesseling, 1994), and anyway there is evidence of net preventive benefits even after any displacement effect is taken into account (Welsh & Farrington, 1999).

There are also concerns that some types of interventions may threaten individual freedoms (e.g., misuse of CCTV and abuse of ID registrations). Research has shown that people are prepared to accept the legitimate use of CCTV surveillance in shops and public places, but are resistant to compulsory identification (Ekblom, 2001). Concerns have also been voiced about the distribution effects of the way different groups or individuals benefit from, or are harmed by, preventive measures (Field, 1993).[2] For example, the undesirable effects of SCP, such as noise pollution from car alarms, can fall on others. There is evidence that such worries have generally receded in recent years (Clarke, 1995), and over all a consensus is forming that situational methods are acceptable in spite of such drawbacks in certain circumstances (Tonry & Farrington, 1995).

High quality demonstration projects and evaluation reviews have shown that when appropriately targeted, designed and implemented, SCP can be a very effective crime reduction measure.[3] For example, a recent evaluation of the predominantly situational measures adopted by businesses since 1990 to control a rising problem of fraud with plastic cards has shown major success (Levi & Handley, 1998). Charge card, check card, credit card and debt card fraud in the U.K. almost halved (from around £120 million) over the period 1991–95, and the ratio of fraud to turnover on sales of goods fell from 0.34% to 0.09%. The mainly situational action against domestic burglary in the Safer Cities Programme is another example of a successful implementation (as distinct from development or demonstration projects) of an SCP program, where action was implemented on a relatively large scale in around 500 schemes over several years. Depending on the background burglary rate, intensity of action, and the outcome measures used, the impact ranged from about 10% to about 30% reduction in expected levels of crime (Welsh & Farrington, 1999; Ekblom et al., 1996).

Whereas the benefits of developmental prevention may be long delayed, SCP has been shown to have immediate benefits, which have been documented in many studies over a long period (Clarke, 1997). SCP interventions are highly focused, so they can exploit the "targetability" of crime problems—repeat victimization, "hot spots," "hot products"—and can bypass intractable social problems that are unresponsive to other approaches. SCP may also have a "multiplier effect" if it prevents crime such as car theft—which is seen as a typical entry to criminal career. Also, if SCP is applied at the design stage of, for example, houses, cars, consumer electronics, and retail electronic point-of-sale systems, it can be very cost-effective in heading off a "harvest of crime." Failure to tackle vulnerability at this stage could bequeath owners, users, and society as a whole, a crime legacy of years in the case of cars, or decades in the case of buildings.[4]

Altogether, most research suggests that Situational Crime Prevention does have enormous potential to reduce the opportunities for crime, and can have an effect in a relatively short time frame. As a result, there is an increasing recognition of the importance of good initial design of products and services for the reduction of crime. Essentially, crime-resistant design is argued to be concerned with making products resistant to the "four misses": misappropriation, or theft; mistreatment, or damage; mishandling of stolen property and counterfeiting; and misuse as resources for crime. The few examples that follow illustrate the "four misses" in terms of both good and bad design:

- For **misappropriation**, a bad design is the old car door-sill locking stud that looked like a mushroom—easy for thieves to get a loop round and pull up. Good design is the tapering version—from mushrooms to asparagus tips perhaps. Another good design—although it may have been serendipitous—was the 110 volt lighting system on London Underground trains. Even though the fluorescent tubes are often exposed, very few are stolen because they are no use to anyone else. This also makes for speedy maintenance because the lights need no special security housing. The opposite problem can happen with car parts such as windscreen wipers. Easy removal on grounds of damage repair and cheap maintenance, combine with high price from the mark-up on replacement parts, to make them a good prospect for theft.

- For **mistreatment**, a familiar bad example is the naïve advertising poster that cries out to be defaced. Products have actually become a lot more reliable in the last few years, so "revenge vandalism" on things like public telephones is perhaps less likely. Provocation can be a side effect of poorly designed preventive measures themselves—but can be avoided.
- For **mishandling**, the era of mass production and disposability means that not only do we have vastly more material goods to desire and to steal, but that they have almost all been anonymous and interchangeable among owners. New methods of production have the potential for increased personalization—who would want to use a mobile phone customized for someone else's ear? In a project supported by the Home Office, synthetic individualization in the form of embedded product identification chips is on the way. Here, where crime prevention is too expensive and bothersome to implement alone, it can hitch a ride on the back of new developments in stock control in the supply chain.
- Offenders are ingenious at **misusing** products for illegal purposes. Until recently, airport luggage labels could be read by any passing criminal—and some made a living from selling information about who was departing on holiday. Now the labels are mostly folded over.

Implementation Issues—How Best to Promote Situational Crime Prevention

In spite of the general consensus regarding the benefits of crime-resistant design, one of the greatest difficulties is getting industry to take crime into account at the design stage of a product or service (Laycock & Tilley, 1995): currently there are no firm ideas about how best to encourage, incentivize or indeed force businesses to design goods and services in the interest of crime control as well as profit. Various ways in which incentives could potentially operate have been put forward, including modifying corporate taxation, altering insurance premiums, shaming through product indices (such as the Car Theft Index), or threatening to withdraw police prosecution resources (Pease, 1998). Currently, there appear to be no studies that have examined the feasibility of such possibilities in detail.

Another related question is who foots the bill for better crime-resistant design. The costs of better design are likely to outweigh the direct benefits to a manufacturer or service provider, in spite of being of significant value to society at large. However this does not necessarily mean that government has to bear all of the costs itself. Pease (1998) argues that governments can often find suitable "interested parties" to fund or facilitate crime prevention—the insurance industry being one such potential collaborator. It has also been suggested that where businesses and other institutions generate opportunities for crime and pay less than an equivalent share of the resulting costs to society, it may be worth considering the "polluter pays" principle. However, this requires the development of a fair and transparent way of tracking and accounting for costs, which may itself be problematic (see Clarke and Newman, this volume, "Modifying Criminogenic Products . . . ").

In sum, a concerted attempt to improve awareness of design against crime in industry and amongst designers, and to encourage design decision makers to take crime-resistant design seriously, would appear to be a valuable and worthwhile project. Nonetheless the as yet unresolved problem remains, how is it possible to encourage industry and designers to design against crime? There is no prior research that has explored how designers currently view the importance of crime-resistant design, nor, apart from earlier research on vehicles (Southall & Ekblom, 1986) is there data on the feasibility for industry to incorporate crime resistance into their product and service development processes. This paper reports the findings of research that sets out to cast light on the current state of crime-resistant design in the U.K., and make some initial recommendations as to how this might be improved.

3. LESSONS FROM THE MILITARY AND FROM ECODESIGN

The purpose of this section is to investigate the approaches used by military designers and those involved in ecodesign, to begin to think about how design has been used to address specific "problems" (the security of military products and equipment, and the environmental impact of products), as well as to identify practices which might be transferable to the area of crime-resistant design. The importance of very clearly defined design criteria is emphasized by the study of military design. The importance of encouraging designers to think in terms of the entire lifecycle of a product is highlighted by the study of ecodesign, as is the importance of creating a climate where addressing the environmental impact of products is perceived by companies to be a source of competitive advantage, rather than a hurdle that has to be overcome.

Military Design[5]

For this part of the study, interviews were carried out and data collected at:

- the Centre for Defense Management and Security Analysis at the Royal Military College of Science at Shrivenham, Wiltshire;
- a company designing defense equipment for operation in land, sea and air based weapons systems; and,
- the Defense Evaluation and Research Agency[6] (DERA), a major U.K.-based defense research agency, which specializes in analyzing defense strategy and scenarios primarily for the Ministry of Defense.

All those interviewed emphasized that security and attack resistance were primary design considerations for military products. Because of the high cost of product loss or misuse, product security is given a high priority and is generally clearly specified in the military design brief. In order to ensure that the highest possible product security standards are met, a carefully defined process, entailing a number of logical steps, is habitually followed. This process entails rigorous problem definition, ensuring that all available solutions and consequences are considered, before finally choosing a particular design solution and then going through a design validation process, to provide an explicit justification for that choice.

For the military designer, capturing all of the details of the product's eventual operating environment is the first, essential task. DERA often plays a critical role at this stage, as it is able to assist designers through the provision of tools and information to which they may not otherwise have access. DERA carries out a variety of different activities, including strategic analysis, detailed system modeling and simulation, materials research and development and weapons research and development. To support these activities, DERA makes use of an enormous range of testing and evaluation resources and techniques, including scenario development, war gaming, computer modeling, optimizing modeling, historical analysis, field trial and exercise analysis, and distributed interactive simulation.

Live testing is an important part of any design process, but for military products it is considered essential to ensure that simulations are as close to reality as possible. Bench testing or stress testing under laboratory conditions can usually only test one or two design elements at once, whereas live testing can stress all elements of the design simultaneously. Moreover live testing is able to take account of human behavior: the physical elements of a design can be replicated in the laboratory, but human behavior under stress is difficult to predict. During live testing any deviations from the expected behavior of the equipment, or human operators, can be observed and analyzed in detail, so as to be accommodated in future designs or design modifications. Indeed, most of the larger defense organizations, such as British Aerospace, have their own test facilities in which they can simulate battlefield conditions and thereby effectively test their equipment under realistic conditions.

Having identified the objectives and the factors impinging on the design, and tested proposed designs under simulated and realistic conditions, the designer must decide on the optimal way to proceed with the design. This is the point in the product development process at which performance targets are compared and trade-offs are made. All designers have to consider factors such as weight, cost, reliability, technical feasibility or avoidance of the loss of primary function in a design. For commercial designers these usually represent important constraints on the final design of a product. For military designers though, factors such as product security and attack resistance are critical. The costs of loss or misuse of a military asset can be extremely high, and therefore the high costs, added weight and so on which might be associated with its protection, are justifiable.

Having made all of the final design choices, there is a further "design validation" process that is routinely followed by military designers, which entails explicitly justifying all of the choices which have been made so that they can, if necessary, be made available to the customer and/or be used to feed into future design and development processes. This is often time-consuming and costly, but it obliges designers to revisit all of the design choices taken throughout what might have been a lengthy product development process, and warrant that these exactly meet the original design objectives. This process often involves verification from outside agencies, including for example DERA.

At this stage, the security and attack resistance of the product is specifically addressed. Manunta (1998) offers a formula which can be used to consider the security of military assets in different circumstances, and which could equally be used to consider the threat from criminals against "civilian" products and services. The formula allows the security of a product to be broken down into three factors for any given situation:

$$\mathbf{S = f(A.T.P)S_i}$$

(S = security, A = Asset, T = Threat, P = Protector, and S_i = Situation)

Although in itself this formula entails little in the way of useful direct prescription about how best to protect an asset, it does provide a means of systematically considering the constituent elements of situations in which the security of military assets is threatened.

In sum, military design practice demonstrates that product security can be given a high level of priority, and it does seem that further study of military design methodologies may be useful in considering how to improve the design of commercial products and services against crime. For example, clear statement of crime resistance as a design priority, the careful use of field data, modeling techniques and attack-testing simulations, and the explicit justification of design decisions in terms of crime resistance may all help to improve the crime resistance of products.

However, there are some of the difficulties that may be involved in designing against crime in non-military environments. For example, because attack resistance is so critical, the military customer is usually able to provide detailed information about the eventual operating environment of the product. Even where there is no specific customer, there is an enormous amount of relevant data readily available, within the manufacturer or at organizations such as DERA, which can help to specify clearly the design criteria related to a military product's security. These data, and the importance attributed to attack resistance, give military designers a significant advantage over their commercial counterparts, ensuring that significant resources and expertise are channeled into this area.

Ecodesign

Ecodesign is an evolving field of study and design practice that focuses on the social and environmental impacts of products and services, and aims to reduce or eliminate these impacts as a result of the output of the design process.

The most common ecodesign initiatives generally focus on *single* environmental issues such as recycling, energy efficiency and design for durability, which are predominantly driven by legislation. However an underpinning philosophy of more sophisticated approaches to ecodesign is the *lifecycle perspective*. The lifecycle shows the journey a product (or system) takes from "cradle to grave" or, as shown in Figure 1, from extraction to disposal. By looking at the whole lifecycle of a product (or service), the environmental impacts of that product (or service) can be attributed to a particular stage(s) of its lifecycle.

The lifecycle framework uses tools such as environmental inventory and impact analysis, and tracks a product in terms of physical outputs across the whole lifecycle. It addresses energy and material flows, transformations of material and disposal of residuals, and in its most complete form evaluates total inputs, outputs and effects for all stages of the lifecycle (Keoleian & Menerey, 1993, p. 12).

It is argued that by taking account of the entire lifecycle of a product or service, it is easier to encourage companies to feel "ownership" of the problems associated with the product/service and to see the long-term impact of their designs, and thus make more informed design decisions. In this way companies are encouraged beyond a single-issue focus, to

Figure 1

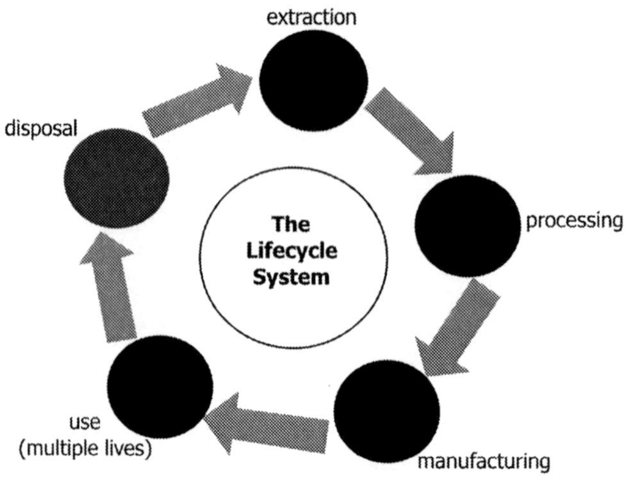

address all the potential environmental impacts that their products may have. This holistic approach can encourage designers to think about the design of their products or services at a number of different levels, each of which has different outcomes:

- product focus—making existing products more resource efficient;
- results focus—producing the same outcome in different ways; and,
- systems focus—questioning the need fulfilled by the object, service or system, and how it is achieved.

Encouraging thinking at these three levels may be as useful for crime-resistant design as it is for ecodesign. A *product* focus may lead to a reduction in the environmental impact of a product, or an increase in its crime resistance, without changing the original product to any great degree. Initiatives are generally single-issue based and result in incremental improvements at specific stages of a product or service's lifecycle. For example, the ecodesigner might add recycled content to a product, look at disassembly issues and improve packaging. A designer concerned with the crime resistance of, say, a public telephone box, might commission better locks and use stronger materials for the cashbox.

A *results* focus questions how the overall function of a product is achieved, and looks to expand current possibilities and provide a wider framework in which possible solutions can be visualized. The function of a product might remain the same but the means of achieving this can result in major product transformation. For example, an ecodesigner might consider that the need for mobility can be better satisfied by an integrated public transport system or a community taxi, rather than a more fuel-efficient car. In the same vein, the crime-resistant designer might consider cashless public telephone boxes rather than progressively more secure measures to protect the cashbox of the public telephone.

A *systems* focus fundamentally questions the reasons behind a product's existence. It looks beyond the narrow boundary of the "produced artifact" and develops new scenarios that may or may not involve the development of a material output. For the ecodesigner, the washing machine may be abandoned in favor of disposable biodegradable clothes after taking into account the costs of water, energy, detergents as well as the machine's materials. For the crime-resistant designer, alternative means of communication or innovative technologies might be considered in favor of producing the public telephone box at all.

As manufacturers move from a reactive to a more proactive position, it has been shown that ecodesign is increasingly seen as a way of achieving competitive advantage through increased customer loyalty, better product innovation, leading the field in technological advances and informing new legislation. Consequently it has become easier to rely on manufacturers seeking competitive opportunities in the area, rather than having to encourage them with legislative and fiscal "sticks." Incremental changes in ecodesign have in the past depended on the latter, especially EU legislation and labeling initiatives (e.g., the energy efficiency of domestic appliances; percentages of recycled material in packaging). However, when companies begin to consider ecodesign as a means of achieving competitive advantage, they begin to adopt sometimes quite radical and innovative approaches (Baynes, 1999).

It is quite possible that by incentivizing designers and industry in general to conceive crime-resistant design in a similar way, it may also engender more innovative outcomes. It could be argued that this already taking place in the field of car design: legislation and some "naming and shaming" of car manufacturers gave the initial impetus to car manufacturers to address vehicle security, but now it is increasingly seen as a measure on which manufacturers can compete against one another. As a result, innovative, cheaper and more effective ways of combating car crime now appear each time new models are released. Experience with ecodesign also suggests that outcomes can be enhanced if designers are inclusive in the design process, engaging with as many actors as possible who are involved at some stage of the lifecycle of the product (or service). Commonly, it is only the initial customer's needs or expectations that are prioritized in the design and development process of a product or service. By promoting a lifecycle perspective, long term impacts of the product or service, as well as "unintended" impacts such as that on crime, are far more likely to be addressed when design decisions are made.

4. EDUCATING AND INCENTIVISING DESIGNERS

The second strand of the "Design Against Crime" project comprised an exploration of the awareness and practice of crime-resistant design amongst design educators and designers, in order to understand how best to improve the capacity and motivation of the latter to design against crime.

To explore crime-related issues in design education, 501 questionnaires were sent to course leaders of the principal design courses listed in the 1999 entry prospectuses of U.K. higher education institutions. Seventy-nine completed questionnaires were returned from thirty-six different institutions[7]—a response rate of 16%.

Forty-one percent of the design educators claimed that crime was not at all relevant to their design discipline, and a further 25% said it was only slightly important. In all cases, environmental, disability and demographic issues were perceived to be of greater relevance than crime. However, there was some variability between design disciplines:[8] 71% of architectural design educators claimed that crime was an important issue for their discipline, whilst almost no fashion design educators thought it was relevant.

The research also showed that with the exception of architectural courses, crime-resistant design is rarely included in design curricula. In the minority of cases where institutions do address crime issues, this is likely to be on postgraduate courses, or as part of a practical project. Despite this barren picture, a majority of design educators agreed that having been made aware of the issue, crime-resistant design was something that could be usefully addressed as part of design courses. Interestingly, following the survey many respondents requested information and literature about crime-resistant design, and commented that the role of design in combating crime had simply not crossed their minds until then. This suggests that raising awareness and disseminating information might encourage more course leaders to include some aspects of crime into their curricula.

In addition, the study did uncover some examples where crime was being effectively incorporated into design curricula. For example, Central Saint Martins College of Art and Design and Nottingham Trent University both made explicit reference to design against crime on their design courses, had effective support for the issue in terms of staff expertise and resources, and were engaged with the issue through student-driven project work as well as staff research. The Design Age Project at the Royal College of Art had also developed crime-related projects with student designers. All three of these examples are relatively recent developments, and might be used as exemplars from which other institutions might learn. Also, their experiences can provide useful information about addressing crime on design courses. For example, one of the institutions acknowledged that it had experienced some difficulty in maintaining student interest in crime-resistant design. Informal discussions with students revealed a widely held view that designing against crime, far from tackling a social issue, is perceived to be driven by insurers and the police. Moreover, it is seen as a "male" design domain, concerned with the design of security locks and devices. With these sorts of attitudes, it was not surprising that there was little interest in the subject amongst students, but this also revealed the need to engage students better with crime as a social problem.

As well as design education, the study explored the extent to which practicing designers were aware of design as a means to combat crime, or indeed took crime into account in the course of their everyday work. Questionnaires were sent to all of the design consultancies listed in *Design Week*[9] Top 100, as well as those listed by the British Design Initiative.[10] In total, 118 companies were approached, and 37 replies received, giving a response rate of 31%.

The results from this survey mirrored the findings from the design education survey. Eighty-seven percent of consultancies stated that consideration of crime had never been a specific client request, and 65% of respondents stated that they or their firm had never designed a product where crime resistance was a design criterion. There were some examples of design projects where designers had specifically addressed crime related issues, but these were often cases where product security was critical, say with street furniture or fraud-prone products or services. Aside from these isolated examples, the general picture that emerged was relatively little understanding of the issues, a lack of useful relevant knowledge and few, if any, incentives, resulting in an overall failure to even consider designing against crime.

These findings were somewhat surprising because in the course of other research considerable relevant knowledge and expertise that was potentially available and of value to designers had been identified. For example, in the police service, Crime Prevention Officers (CPOs) are charged with disseminating relevant information about crime and criminal behavior, whilst Architectural Liaison Officers have specialist insights and understanding of how good architectural design decisions can be made to reduce crime in the future. There are also comprehensive information resources about crime-resistant design available from the Crime Reduction College, the Home Office as well other sources such as the security industry. Nonetheless the research showed that designers were either unaware or skeptical about the value or relevance of these resources. For example, Architectural Liaison Officers were invariably consulted in the later stages of a building project, when many design decisions had been fixed, as they were perceived not to make any positive contributions to the overall design process.

In sum, the research suggests that (apart from architects) there is no systematic education of designers to consider the potential of good design to combat crime, and where designers are aware of the issues it usually comes about as a result of *ad hoc* projects where product or service security has been a critical design criterion. This results in designers having an image of designing against crime that is largely negative, constraining and compromising, reinforced greatly by general ignorance of the information, data and methods that would enable them to apply it to their practice.

Recommendations

The findings point to the need to raise the profile of crime-resistant design amongst design educators, and to incentivize designers to design against crime as a matter of course in their everyday work.

The embedding of crime issues into design courses at both degree and school levels could provide a powerful tool for raising crime awareness generally, and ensuring that designing against crime becomes part of everyday design best practice. While teaching design against crime is currently a marginal activity that has yet to capture the imaginations of students or tutors, the inclusion of other topics in design education, such as environmental and disability issues, has shown that with effective support and encouragement, the design curriculum has the flexibility to adapt and apply new disciplinary approaches within its core curriculum over a relatively short period.

The experience of incorporating social issues into design courses also underlines the need for integration of new subject matter into the core design curriculum, lest students view it as being of marginal interest. A key long-term challenge will be to ensure that design against crime becomes a central component of design courses, rather than a course option, or an issue which is only addressed as an element of project work—whilst the emphasis on project work enables design courses to respond flexibly to new issues, the research suggests that other teaching and learning methods are required to provide students with a firm grounding in specific subject areas. Another challenge will be to persuade students that crime-resistant design tackles important social problems. Possibilities which might help here include commissioning inspiring exhibitions and content-rich websites, getting police forces to offer work placements to design students, and arranging for students to apply their skills on community projects addressing issues such as vandalism and street crime.

The research suggests that it will not be sufficient only to promote design against crime at the initial design education level; it also needs to be better incorporated into on-going education and training of practicing designers. It is anticipated that there will be little resistance to promoting design against crime amongst practicing designers, as it seems to present a number of distinct opportunities to the British design industry. In particular it would seem to provide Britain's £12 billion design consultancy industry with a critical competitive edge in a vigorously fought international market: developing a specialist expertise in this field would strengthen Britain's prospects as a design consultancy exporter, which already accrues annual export earnings worth over £350 million.[11]

One way to raise awareness and focus resources would be to make use of existing design bodies such as the Design Council, the Design Business Association, British Design and Art Direction, the Chartered Society of Designers and the Royal Society of Arts. These key organizations have a significant influence within the industry: 80% of the designers surveyed stated that they where members of such organizations. During the course of the research, representatives of all of these bodies were interviewed to ascertain their attitudes to design against crime, and their willingness to become involved in promoting the issue within the design profession. Many of those interviewed agreed that the design profession currently regards crime prevention in a negative way. However, all of the design organizations consulted offered to support future initiatives targeted at practitioners, educators and students. Indeed some have already recognized the value of designing against crime and have introduced their own initiatives. For example, the Royal Society of Arts (RSA) has introduced design against crime briefs as part of its annual bursary award program for design students, and claimed to be learning a lot about what does and does not appeal to students, what information tutors require, how easy it is to find relevant information, and so on.

Several other suggestions were made by design organizations about how they could contribute to the promotion of crime-resistant design amongst designers. These included awards and competitions, an annual exhibition and conference showcasing the best of design against crime and telling the development stories behind them, profiling the issue in *Design Week*, and a "Design against Crime Kitemark" awarded to products, services and environments where crime issues had been most effectively and creatively addressed in the design process. Some organizations also recognized that currently there is no single comprehensive source of design against crime information provided by any one organization, as most design associations have a specific constituency and method of working. There are opportunities to construct a single on-line information source with relevant hyperlinks to the various different design bodies, case study material, data on crime and research papers.

In sum, although design against crime was generally not part of the existing designers' or design educators' repertoires, the research revealed some enthusiasm among those in charge of design course for promoting the issue, so that it becomes an everyday part of design practice. It seems that the mechanisms and processes exist that would allow design against crime to be promoted relatively easily, in order that designers are better educated and incentivized to design against crime as a matter of course. With some directed effort, government could make use of existing channels, for example organizations such as the Design Council, to tap into the broad enthusiasm for the idea that, alongside a reputation for its inventiveness, quality and attention to detail, British design could also excel in its capacity and willingness to design against crime.

5. PERSUADING AND ASSISTING INDUSTRY

Having briefly recounted some of the findings related to the education and incentivization of designers, the next step is to examine the practicalities of implementing crime-resistant design in business. It would have been impossible to survey every type of manufacturer or service provider across the spectrum of British industry, so instead six sectors were selected (automotive, consumer electronics, railway rolling stock, schools, built-environment and e-commerce) where a range of different issues could be investigated. In each of these sectors a range of companies was studied—in addition to Trade Associations, training bodies and standards-setting agencies—to identify and explore the real-world factors that enable or constrain businesses' ability to design against crime.

Although focused on only six sectors, the research illustrates the tremendous variety and complexity of design and development processes not only in different industries, but also in different companies in the same industry. Consequently, there are a myriad of different design decision makers, enablers, constraints, and design tools and techniques which have to be considered when thinking about how companies might be encouraged to design against crime. In spite of this variety, it is possible to discern some common factors that determine whether crime resistance is incorporated into a company's new product development process. It was found that even if there is good awareness of crime issues amongst designers, there are a number of practical issues that either facilitate or constrain the incorporation of crime-resistant design in the development of new products or services. However, by supporting industry to develop appropriate resources and communication channels, and persuading them to make use of them by way of appropriate legislation and consumer pressure, is feasible to create the conditions where organizations are likely to design against crime as a matter of course.

Capacities to Design against Crime

One of the most important factors found to prevent the incorporation of crime resistance into new product/service design, was the complexity and fragmented nature of most modern product/service development processes. The majority of manufactured goods and services are nowadays extremely complicated, and the notion that there is a single designer responsible for the overall design is a gross over-simplification. Moreover, many of the "design decision makers" may not themselves be designers, so even the best design training in the world may not be a ultimately have a strong effect on the crime resistance of the finished product or service. Also, "design" is rarely an activity carried out under a single roof: frequently, many of the parties involved in making design decisions are based in different countries, so it is unlikely that crime considerations of any one particular country will figure highly in any final design. In this context, it is relatively difficult for crime resistance to be championed unless it is given a very high priority in the development process.

For example, respondents in the railways industry confirmed that crime resistance is recognized as a relatively important issue in the development of new train carriages. However, there are many different parties involved in the design and development of rolling stock, who are often based in several different countries, including the Train Operating Companies (TOCs), carriage manufacturers, equipment and fitting providers and various component manufacturers. In most of the cases examined, crime resistance was subordinated to other design priorities principally as there were no effective mechanisms through which to exchange relevant knowledge and coordinate design decisions between these various parties.

Complexity was not an intractable problem, though, where crime resistance was given a high priority. For example, the new product development processes in the house building and automotive industries were also shown to be complex and fragmented, but had in place systems which enabled this complexity to be overcome. In the case of house building, it is the local planning process which serves to ensure that design decisions can be coordinated and information shared among property developers, architects, local builders, council planners and the police (although as will be discussed later, there are other issues which sometimes prevent crime resistance being implemented). As for the automotive industry, a finished car may include components from hundreds of individual manufacturers, which are themselves the result of hundreds of different "design decisions." However, because of the importance attributed to crime resistance in the industry (to be discussed later), it was found that sophisticated coordinating mechanisms and communication procedures had been implemented to ensure that, in what is an extremely complex product development process, crime resistance is effectively addressed by all relevant design decision makers.

Another constraint on the incorporation of crime resistance features into new products and services was the perceived priority of other design goals, like practicality, health and safety, aesthetics or low cost. For example in the car, consumer electronics and e-commerce sectors, companies were reluctant to introduce any feature that would render their products difficult to use. Architects, on the other hand, claimed that the aesthetics of a design would take precedence over considerations about crime resistance. In some simple cases, it was claimed that "good" designers could sometimes synthesize these conflicting design goals. However, with most complex products, especially where there are a large number of design decision makers, it was acknowledged that arriving at such a synthesis was often impossible, in which case the various different criteria would be prioritized: crime resistance would invariably feature towards the bottom of the list. There was only one sector where crime resistance (or "security" as it was usually referred to) was perceived to be the most important design criteria—schools. Here, crime-resistant design was generally given very high priority, but even so respondents claimed that they often had to make compromises given the competing concern with cost.

Indeed, in all of the sectors explored, low cost was uniformly given as one of the most important design criteria for any new product or service, and this was felt to conflict with any requirement to design against crime in several ways. For example, it was often claimed that crime-resistant products or services invariably required additional components or processes which added to the cost of production or delivery: cars may require additional electronic circuits and stronger materials, and houses may need better quality locks, alarms and extra lighting. In addition, designing against crime was often felt to necessitate additional research and development costs, especially where technical innovation is required to improve crime resistance. Also, and perhaps less obviously, it was argued that crime-resistant features often have to address local or national crime problems, which conflicted with many companies' drive for economies of scale. For example, consumer electronics and car companies favor a standardized product for global markets to achieve maximum economies of scale; adding features to address U.K. specific crime problems may not be seen as particularly attractive. House building also favors standard materials, design and equipment, and the cost of adapting the standard product to account for local crime problems may be very costly.

The third key factor that had the potential to constrain a company or industry from effectively designing against crime was the lack of, or poor utilization of, relevant tools and resources, in particular crime data and information about design "best-practice." Most sectors did not share information about crime-resistant design, and many companies were unable to report where they might find information that was relevant to their particular company or sector. For example, none of the Train Operating Companies appeared to be making use of crime data from the British Transport Police to inform the design decisions being made about new rolling stock. In the building industry, although sophisticated software is used extensively by architects and planners in the planning of new building developments, with which architects can model sight lines, the effect of different lighting and so on, it is was rarely used explicitly to address crime related issues. Moreover there is also a large pool of relevant knowledge and data on crime-resistant architecture, but the study showed that little or no use is made of the local crime data, and police Architectural Liaison Officers are often only consulted late in the design process of a new building development.

However, in the automotive and e-commerce sectors, tools and resources were used extremely effectively to design against crime, which served to emphasize the potential for other sectors to do likewise. For example, the automotive industry has developed highly sophisticated simulation tools to allow designers to build "virtual vehicles," which are used to analyze the performance and interaction of components, including security and other crime-resistant features, without incurring significant cost. Also the Motor Insurance Repair Research Centre (MIRRC) and the Society of Motor Manufacturers and Traders (SMMT) collect broad data on vehicle crime, which they pass on to individual manufacturers. These organizations also employ specialist engineers who are available to advise manufacturers on crime-related issues throughout the design and development of a new vehicle, and administer security assessment protocols that allow the performance (including crime resistance) of a new vehicle to be tested under real and simulated operating conditions. To determine how the systems will perform prior to these assessments, vehicle manufacturers are also able to draw on a wealth of other outside contractors (including some set up by former car thieves) who can provide real-life attack testing.

Within the e-commerce sector, the use of simulation and analysis to help design against crime is completely integrated and fundamental to the development process. There are a large number of "off-the-shelf" software systems that designers use to provide and test security within their system. Also expertise and software resources are available which can simulate and analyze the typical use and abuse that these systems will be subjected to once launched. Then, there are a number of external accreditation schemes that organizations can use to evaluate the security level of their services, including commercial schemes such as Verisign, or government-led initiatives such as the Information Technology Security Evaluation Criteria (ITSEC) scheme in the U.K. Once e-commerce services are launched, the process of testing and monitoring usually continues, to ensure that any unforeseen problems are detected as quickly as possible.

Drivers of Design against Crime

While all the factors discussed thus far may render it more or less difficult for an organization to achieve crime-resistant design solutions, they typically do not *determine* the design decision-maker's motivation to adopt a crime-resistant design solution in the first place. Put differently, organizational complexity, design trade-offs and the availability/use of tools and resources are factors which enable or constrain an organization's capacity to design against crime once the decision has been taken to do so, but in themselves they only determine the decision to develop a crime-resistant design solution to a limited extent.

The research suggests that there are two external factors that constitute the primary drivers of decisions to develop crime-resistant design solutions, namely the *level of public regulation* and the *degree of consumer pressure*. In cases where the level of public regulation and/or consumer pressure are strong (such as in the car industry or e-commerce), companies are forced to adopt some form or crime-resistant design solution, independent of the internal enablers and constraints, even where it may be complex or costly. On the other hand, in cases where consumer pressure and the level of public regulation are low (such as in consumer electronics), companies tend to fail to adopt crime-resistant designs. As such, any effective strategy that attempts to enforce the adoption of crime-resistant design solutions has to focus primarily on these two external drivers.

That the automotive industry now designs against crime as a matter of course, was initially brought about in part by the underlying requirements of legislation that govern the inclusion of security devices within cars. In a design trade-off situation in the automotive industry, the research showed that elements that are subject to legislation are now given high priority, followed closely by those that influence insurance requirements. The rail industry, by contrast, is very closely regulated with respect to crash worthiness and safety, but not closely regulated as far as crime related issues are concerned. As a result, TOCs do make some efforts to design against crime, but not in the systematic way that the automotive industry does. For example, there were examples of TOCs using market research tools to discover the needs and wishes of their customers, where crime was usually addressed. However these surveys were usually undertaken on an *ad hoc* basis, and relevant information on crime-related issues was not always fed back to the manufacturers of the rolling stock effectively (in contrast to issues relating to train safety and crash worthiness).

By itself however, legislation did not appear to be especially effective in creating the conditions for companies to pro-actively design against crime. This was most clearly demonstrated in the building industry. For example, various legislation, standards and guidelines including building regulations, local planning requirements, "Secured by Design," BS8820 and so on, variously address the crime resistance of new building developments. Yet given the absence of any central, monitoring body, the opportunity for local interpretation of guidelines, and the complexity of many standards, all of the various parties interviewed acknowledged that rules were rarely followed to the letter. All confirmed that enough would be done to comply with the basic criteria of any legislation or standards, but these would rarely be exceeded.

Although regulation seemed to provide basic minimum standards for crime-resistant design in some of the sectors examined, it was found that ultimately it is consumer pressure that is the most important factor in creating the conditions where crime-resistant design is taken seriously and acted upon creatively by organizations. For example, although initially driven by legislation, the enormous public attention which has been focused on the crime resistance of cars by government and consumer organizations has led to strong impetus within the automotive industry for better security and crime resistance. In the e-commerce sector, the perceived sensitivity of consumers to security issues seems to have been critical in ensuring that crime-resistant processes are paramount in the provision of services: if security is compromised, the reputation of an e-commerce organization evaporates, as do its customers.

By contrast, the lack of strong consumer pressure appears to be an important determinant of the low level of crime-resistant design in industries like consumer electronics, housing, and train carriages. For example, the consumer electronics sector has the expertise and resources to design and implement security features for its products, and yet crime resistance is perceived to be a relatively unimportant product criterion amongst consumers, so it is generally ignored. As discussed earlier, in the building industry there are also good tools and knowledge available, yet these are rarely used to their full potential. The same is true of the rail industry, where there is good relevant information collected by the British Transport Police amongst others, which is rarely used by the TOCs when it comes to designing new railway carriages. In all of these cases, the research indicates that without a strong perceived demand from consumers, designing against crime was seen as a fairly low priority.

The study suggests that the level of consumer pressure exerted on a particular industry or company depends on two factors: the initial level of consumer awareness and interest in crime resistance, and more significantly, the extent to which there are mechanisms which allow this interest to be channeled towards the relevant design decision makers. Consequently, although there is some slight variation in the awareness and concern of consumers with the security of houses, schools, cars, trains, consumer electronics and e-commerce, there are significant differences in the way that this awareness and concern is directed towards those who are able to influence design decisions in these different sectors. For example, although there seems to be a stronger concern about the security of houses than e-commerce services, less direct pressure seems to be felt by housing developers than e-commerce companies. How is this accounted for?

One key determinant may be the cost of switching between different providers of the product or service. In industries like e-commerce where the switching costs are low, not least because the consumer typically has a non-Internet alternative at his or her disposal, the service provider knows that consumer concern about security readily translates into the loss of reputation, customer and revenues. With houses on the other hand, consumers find it extremely costly to switch to another product (i.e., move house) in order to address a problem of low crime resistance.

Another factor in translating concern into pressure is the extent to which a manufacturer or service provider is clearly identifiable as bearing the main responsibility for implementing a poor design, which subsequently facilitates crime-related problems. For example, housing development companies are unlikely to be identified by consumers as having primary responsibility for the crime resistance of new housing developments—there are numerous other companies, bodies and factors involved, which might also be held responsible by residents (or blamed by other organizations) for having allowed or implemented poor design features. Moreover, once a house is built the property development company tends to have very little contact with the purchaser, in contrast to, say, the local council, which as a result of its on-going proximity to residents is more likely to bear the brunt of any criticism. Consumer electronics companies are also generally insulated from crime-related effects or outcomes associated with their products. Consumers are more likely to attribute crime to other factors, such as security of the home or protection offered by the owner, rather than the poor crime resistance of the product. Also, insurance serves to insulate manufacturers from negative feedback from consumers; indeed, as it is likely that a stolen or damaged product will need to be replaced (on a new-for-old basis), both manufacturer and consumer may benefit from the poor crime resistance of the product.

Schools present a similarly complex picture, with multiple stakeholders (Local Education Authorities, builders, central government, head teacher, governors, caretakers) who might be held responsible for design decisions that facilitate or prevent crime. Yet here far clearer "ownership" of the problem by the head teacher and governors was found. The research suggests that they take on this responsibility as they are affected and have to deal with crime problems associated with any bad design decisions, and have to face their "consumers" (pupils and their parents) to explain and solve any crime-related problems on a daily basis. However the most instructive example, showing how consumer pressure can be effectively channeled to bring about changes in thinking about crime-resistant design, is the case of the automotive industry. As already discussed, the automotive industry has complex design and development processes, which involve a myriad of different parties making design decisions that can impact on the ultimate crime resistance of a product. However consumer pressure has been effectively focused on the car manufacturers by high-profile product indices ranking the crime resistance of different models (for example the Car Theft Index, discussed earlier), and by differential insurance premiums that to some extent also reflect their crime resistance. As a result, car manufacturers do their utmost to ensure that all stakeholders understand the importance of designing a vehicle that is as far as possible, crime resistant.

In sum, a key finding of the study is that if a company feels direct pressure from the public to address the crime resistance of their product/ service (or indeed if the company anticipates the potential for such pressure), it will tend to design against crime effectively. Therefore the principal recommendation is that consumer awareness of product/service crime resistance and security must be raised, and then mechanisms and processes must be established to translate this awareness into pressure on manufacturers and service providers.

6. INFLUENCING CONSUMER ATTITUDES

Having suggested that consumer pressure may be a key factor in encouraging companies and designers to design against crime, this section reports the findings from focus-group and in-depth interview research which was carried out to ascertain the general public's levels of awareness and attitudes to the crime-resistant properties of products and services. In order to provide useful data that could be linked to findings in the rest of the research project, this consumer research focused on the crime resistance of cars, consumer electronics, train carriages, e-commerce, the built-environment and schools. The research was undertaken by a combination of focus group discussions and in-depth face-to-face interviews in the South of England, the Midlands and the North of England, with consumers selected to represent the range of socio-economic groups, ages and genders.

The focus-group research began by asking respondents to think about their purchase criteria for cars, hi-fi equipment and houses. The criteria that were spontaneously mentioned by respondents did not include crime resistance, except in the case of house purchase, where the "safety of the area" was said to be important. Each of the factors they had cited then were listed, "crime resistance" was then added to the list, and respondents were asked to rank these on a scale of 1–10 (10 being most important). Crime resistance was shown to be the most difficult criterion for respondents to rank: whereas criteria such as price, reliability, features, color, make and model were given rankings that were broadly consistent amongst all respondents, the ranking given to crime resistance was the most variable and also, on average, the lowest.

Next, respondents' opinions about crime in general were solicited. There was a strong feeling amongst many of the more vocal respondents that more could and should be done about crime, and there were strong opinions that the best way to tackle crime generally. Most discussed the importance of addressing what they called the "root causes" of crime, including drugs, unemployment, poverty and lack of facilities for the young. Also it was considered critical to be tougher on criminals, by increasing sentences and making prison conditions harsher. Many also mentioned the need to increase the number of police and/or security guards on the streets.

Situational Crime Prevention did not initially figure in any of the discussions. However, once initial opinions had been expressed, respondents were asked to consider different types of crime that concerned them. In each focus group, a distinction was quickly made between opportunist acts and those carried out by habitual offenders. There was greater acknowledgment of and concern about opportunistic crime amongst lower socio-economic groups than among the higher ones, which was usually attributed to the fact that the former often live in estates/areas where opportunist crime was felt to be common. After making the distinction between habitual and opportunist crime, all groups acknowledged that much could be done to reduce opportunist crime. The main, unprompted suggestions included better locks, alarms and immobilizers on cars, better CCTV coverage of public areas, especially schools, trains, stations and car parks, better planned housing estates and better lighting.

When questioned directly about the crime resistance of consumer electronics, which was never mentioned spontaneously by respondents, it was commonly felt that this was inappropriate. Respondents commented that consumer electronics were of relatively low value which depreciated quickly, they were easily replaced if stolen, were covered by insurance (moreover this was often "new for old" insurance), and there was usually no sentimental value attached to the products. Indeed many respondents said that if an electrical item was stolen, they would not want it to be returned as they felt that it had been "tainted": consequently, there was strong resistance to design features that were aimed at returning electronic goods. Also respondents worried about the conflict of crime-resistant features with the product's ease-of-use. For example PIN numbers to operate the electronic item were hard to remember as well as difficult for small children/elderly to operate. One respondent summed up the general feeling about the value of designing against crime for consumer electronics as follows: "A video is a video, a TV is a TV [. . .] there are more important things."

When respondents' attitudes to the crime-resistant design of houses were explored in more detail, there was a quite different reaction. As discussed earlier, the house was the one"product" for which security was spontaneously discussed as being an important purchase criterion. In exploring why this was felt to be different from the other products and services discussed, respondents invariably reported that this was because it involved a different type of crime, crime against the person. Many respondents elaborated by saying that they strongly feared personal attack, whereas theft of items was perceived as more of an annoyance. In no cases did it seem to cross the mind of respondents that it was the ease with which consumer electronics could be stolen and distributed that was the principal factor driving the burglary of their homes. One respondent neatly summed up the general attitude of respondents, implying that efforts targeted at improving home security were more important than those aimed at protecting consumer goods: "The violation would be in the entry and not in the stereo."

In spite of some broad agreement that better design was a common-sense and potentially valuable means of reducing certain types of crime, the research revealed a clear underlying resignation about crime in general. The majority of respondents believed that there is a certain "level of crime" that is fixed and will be maintained, regardless of any crime reduction strategies that are implemented. For example, many respondents, especially those living in the areas with high levels of crime, claimed that thieves would always be one step ahead of those responsible for improving security: "There are always some things that they can break into, whether they've got alarms or not, they'll just keep trying until they do. There's a way to do everything."

But interestingly, many respondents also said one of the most significant ways of reducing crime was for them to act more responsibly themselves. Measures that many respondents suggested included:

- increasing security around the home;
- being careful to lock doors (of cars and houses), not leaving valuables on show; and,
- participating in community initiatives such as tenants' associations, parents' associations, neighborhood watch and so on in order to put pressure on local authorities and landlords to improve lighting, layout, and security features.

Such commonly-held attitudes suggested that in spite of some misgivings, there is an implicit recognition that by altering the situation, opportunist crime can be reduced.

It should be added that this research on awareness of and attitudes to crime-resistant design was exploratory, and requires further quantitative studies with larger samples in order to provide more detailed data. However, broadly speaking, it was found that crime-resistant design is not currently held by the general public to be the most important crime reduction strategy. This suggests that if consumers are to exert meaningful pressure on manufacturers and service providers, Situational Crime Prevention needs to be presented as a measure that can complement other "tougher" crime prevention strategies, not substitute for them. Also, the research suggests that efforts to raise awareness of the benefits of better design against crime should be targeted in certain sectors, where better design is already considered to be a potentially useful and commonsensical initiative (for example, home security, the built-environment and public transport). Finally, it seemed that the advantages of crime-resistant design were better appreciated in areas which had received attention in the media, such as cars and CCTV in public areas: providing the media with good information seems to be important means of persuading the public of the benefits.

7. RECOMMENDATIONS FROM THE MAIN REPORT

At the beginning of this paper it was argued that although Situational Crime Prevention is known to be a potentially valuable and cost-effective crime reduction strategy, the question of how to encourage manufacturers and service providers to design against crime is currently poorly understood. This paper has summarized the empirical research that teams at the Universities of Cambridge, Sheffield Hallam and Salford have carried out to move our understanding forward. It was found that with some exceptions, the training and awareness of designers about crime-resistant design is currently poor, and design against crime in industry is generally undertaken on an ad hoc basis. Although there are some examples of effective design against crime, in most design disciplines, industries, and companies, the potential for good design to help reduce crime is simply not realized.

Nonetheless, it has been argued that there is considerable potential to educate and incentivize designers, and persuade and assist industry to design against crime as a matter of course. This final section summarizes the various recommendations made in the main report, which hopefully will encourage and facilitate better design against crime in the U.K.

Training

The first recommendation is to raise awareness and improve the training of designers. These various initiatives aim to embed Design Against Crime within design curricula at school, degree and professional levels, promote the development of teaching resources and foster a community of informed, committed educators. The research showed that investing only in initial design education is likely to be insufficient in itself, and therefore in-service training of practicing designers should also be addressed. The various existing design organizations would be valuable partners in driving this initiative forward. However, the study also warned against focusing resources only on "designers." The empirical cases explored showed that in many cases, the important design decision makers were not always "designers": they could equally be engineers, marketers, financiers or even head teachers. Raising awareness of design against crime amongst trade organizations might prove a useful means of reaching these less obvious "design decision makers" and convincing them of the benefits of designing against crime.

Information and Resources

The second recommendation is to make better use of available resources and data on crime and crime-resistant design issues. The study has shown that there is a wide variety of relevant data already available on crime and crime design, but this is disparate and often difficult for companies to access. Moreover, for data to become useful *information*, it needs to be interpreted. Often this does not happen. In some cases, it may be that the data that would be useful for designers or companies to feed into their design decisions has been collected in a way that renders them inappropriate, or are held in a way that makes them difficult for design decision makers to interpret. For example, police Architectural Liaison Officers are an invaluable source of information for the building industry, but they are often not consulted effectively. One reason for this is that their advice tends to be specific to a locality, whilst key design decisions tend to be made at a national level by large property development companies. Capturing local data in a national database might be one possible solution. There are many other cases where the collection, interpretation and dissemination of relevant crime information could be improved by considering how it is best accessed and used in practice. Encouraging greater collaboration between designers, industry and commerce, crime prevention agencies and academic researchers would certainly go some way to improving appropriate information flows and developing relevant knowledge bases.

Trade and Design Associations

It would seem beneficial to make better use of trade and design associations to accumulate and disseminate information about crime-resistant design. The study found the accumulation and dissemination of knowledge and best practice to be very good in some industries, but non-existent in others, which often seems to be associated with the strength or weakness of industry-wide associations. The MIRRC at Thatcham is a good example of an organization that serves to coordinate and disseminate knowledge about security and crime resistance throughout the automotive industry. MIRRC is able to provide information, resources, skills, testing facilities, cases of best practice and so on, which are not easily available or are costly at the individual company level. It is recommended that such associations are identified and encouraged to act as industry-wide coordinators of design against crime knowledge and best practice. As well as intra-industry knowledge, the research suggested that there might well be opportunities for companies in one industry to learn about best practice in other industries. For example, many interviewees in the railway sector identified the automotive and aerospace industries as useful role models for learning about the security and crime resistance of railway rolling stock. This inter-industry transfer of knowledge and might usefully be encouraged. Here design associations may be the more appropriate bodies to use to promote cross-industry or cross-discipline learning.

Regulation

Codes of practice, standards and regulation appear to be potentially powerful means of encouraging crime-resistant design to be incorporated by companies, and it is recommended that in cases where consumer pressure is absent or embryonic, such approaches be considered. However, it appears from many of the cases examined, that companies are likely to respond to regulation by simply doing the minimum to comply. Also, in many cases, the international scope of a company's operations may make U.K.-specific regulation impractical. For example, the majority of consumer electronics, cars and railway carriages are designed outside of the U.K. for global markets, and it is unlikely that these companies would warm to regulation being imposed that addresses U.K.-specific crime problems.

Consumer Pressure

The final recommendation report makes is to find ways to facilitate the channeling of consumer pressure on manufacturers and service providers. Public awareness and attitudes to crime-resistant design were discussed in section 6, and although it is seen as less important than an effective criminal justice system, it does appear that there is a latent feeling amongst the general public that better crime-resistant design can be effective in reducing some types of crime. This suggests that any initiative to strengthen design against crime must be shown to be a complementary approach to crime reduction, rather than a new stand-alone strategy.

There are several ways to increase awareness and influence attitudes to crime-resistant design. One of the most powerful means that can be inferred from the survey of public awareness and attitudes is to promote the advantages of crime-resistant design in the media, especially television: respondents frequently had positive attitudes to good crime-resistant design that they had seen on television or in the newspapers. Another means of channeling public concern so that manufacturers and service providers pay attention (which again seems to have been fairly successful in the past) is the publication of indices that rank the crime resistance of various products. The consumers who were interviewed as part of this study were aware of the existence of such rankings for cars, and more importantly, so were the manufacturers. Here consumer associations and insurance companies might be useful allies in raising the awareness of product/service security amongst the general public, which would serve to put pressure on manufacturers to design against crime in the first place.

Putting It Together

All of the recommendations suggested so far have the potential to improve significantly the extent to which products and services are designed against crime. However, a piecemeal strategy, addressing one or two of the key issues that have been identified in isolation, is unlikely to create the conditions where crime-resistant design becomes naturally embedded in product and service development. Although this paper has tried to focus on the broad issues which serve to enable or constrain design against crime, hopefully it has also conveyed the heterogeneity of enablers and constraints, tools, techniques and resources, incentives and disincentives, all of which contribute to particular design disciplines', industries' and companies' willingness or ability to design against crime.

One way of addressing this problem would be to establish a permanent body to co-ordinate "Design Against Crime" in the U.K. A permanently established organization could target industries where crime-resistant design would be most effective, could channel consumer pressure where needed, could suggest codes of best practice or where necessary lobby for standards or regulation, could anticipate future trends in crime and technology, and so on. Such a body would be able to ensure that the *capacities* here discussed are improved, but moreover could have a significant influence on ensuring that the *drivers* discussed are operating effectively. The principal roles of such a body might include the following:

- **Coordinating and effectively disseminating relevant information and knowledge about crime and crime-resistant design.** This would entail gathering together relevant crime-related data in one place, commissioning research and reports to provide more specific data where needed, interpreting data so that it is accessible to design decision makers, and coordinating efforts and acting as a central resource for existing design and trade associations. In particular, intelligent use of new technology could ensure that relevant information is not just made available, but is directly relevant to the needs of companies.

- **Targeting industries and companies where effective crime-resistant design is needed, identifying key blockages and taking action to overcome them.** Both crime and technology are dynamic, and as such a permanently established body could ensure that available resources are always targeted at areas where they are most cost-effective.
- **Focusing public attention, so that manufacturers feel that crime resistance is appreciated as an important product attribute.** Ways of raising consumer awareness of and interest in the crime resistance of products might be affected through the compilation of indices of more or less crime-resistant products. Also recognition of well-designed products would seem to be valuable: in this context it would be better to champion good products and services, rather than "name and shame" bad products and services, as this would tend to motivate companies to value crime resistance as a marketable product attribute.
- **Anticipating future crime trends and the crime-reducing potentials of new technologies.** An important advantage of a permanently established body is that it would be able not just to act on existing information, but could act as a repository of expertise, and a forum for experts in the field, in order to anticipate crimes which might be mitigated through better or more carefully considered design.

There is a possible danger that both the public and companies might interpret the establishment of such a body as government being both "soft" on crime, and intrusive or interfering in the commercial world. However to overcome such potential criticism, a possible model to consider might be Groundwork, a charitable trust set up by what was then the Department of the Environment in 1981 to focus on environmental regeneration. Groundwork works in partnership with local people, local authorities and business to promote economic and social regeneration by improvements to the local environment. It is independent, but receives core funding from the Department of the Environment, Transport and the Regions. It seems to be viewed very positively by all the parties it works with, being perceived as a valuable, independent resource which is able to advise, assist and co-ordinate the efforts of companies, local authorities and local communities to address local environmental issues.

An equivalent independent body which focused exclusively on the crime resistance of products and services, could channel consumer pressure, could suggest codes of best practice, or where necessary lobby for standards or regulation. It could work with industry as a source of advice and information, and in this role could probably generate revenue to help cover its costs. It is possible that such a body could multiply funding received from central government by generating revenue through for example testing facilities, consulting advice and dissemination of information. Such a body could potentially be established at either a U.K. or an EU level—here it would be necessary to weigh up, for instance, the advantages of being able to focus on U.K.-specific crime issues against the advantages of having Europe-wide backing when negotiating with large multinational companies. A detailed feasibility study would be required to address these and other issues, but on the basis of this empirical study, the possibility of establishing such a body should be given careful consideration.

Address for correspondence: Simon Learmount, Judge Institute of Management, Trumpington Street, Cambridge CB2 1AG, United Kingdom.

Acknowledgments: The research for this paper was carried out at the Centre for Product Innovation Research, University of Cambridge, the Design and Innovation Research Unit at Sheffield Hallam University, and the Design and Innovation Group at the University of Salford. A full account of the research and its findings can be found in the report, "Design Against Crime" (September 2000, Cambridge University Publications Centre), available at www.shu.ac.uk/schools/cs/cri/adrc/dac/designagainstcrimereport.pdf

NOTES

[1]Details of the U.K. Car Theft Index can be found at: http://www.crimereduction.gov.uk/cti2k.htm

[2]For instance, a laissez-faire approach can lead to inequity, with displacement of crime from the better off to those less able to protect themselves or obtain publicly funded resources.

[3]Ekblom (2001) suggests that SCP measures can lead to a 40–70% reduction in expected levels of crime.

[4]Evidence for the effectiveness of government-orchestrated local action at the design stage comes from a Dutch housing estate scheme (Ekblom, 2001). For approval, developers' projects must meet standards covering residents' participation, neighborhood management, home watch, and building design including layout of rooms and entrances and target hardening. An evaluation in Rotterdam showed a 70% reduction in burglaries after one year between those new houses involved in the scheme, and those not.

[5]While the present study focuses on the basic design process, Ekblom (1999) explores specific design against crime lessons from military arms races and natural equivalents.

[6]Formerly the Defense Research Agency (DRA).

[7]Fifteen responses from visual communication courses, 14 from architecture courses, 12 from fashion courses, 10 from product/industry design courses, 28 from various other design courses.

[8]For the purpose of this study, the multitude of different design disciplines have been broken down into 4 categories: visual communications, architecture, fashion and industrial/product design.

[9]*Design Week* is a trade journal that is read by many designers and reflects current design issues and profiles current design projects across all design disciplines.

[10]This organization promotes U.K. design globally, and claims to list details of the leading British design consultancies.

[11]Source: Department of Culture Media and Sport (1998), Creative Industries Mapping Document.

REFERENCES

Baynes, A. (1999, October 28). *Ecodesign: A practical corporate approach*. Environment Council DfE Seminar. London.

Clarke, R. V. (1995). Situational crime prevention. In M. Tonry & D. P. Farrington (Eds.), *Building a safer society: Strategic approaches to crime prevention*. Crime and Justice, A Review of Research, Vol. 19. Chicago: University of Chicago Press.

Clarke, R. V. (Ed.). (1997). *Situational crime prevention: Successful case studies* (2nd ed.). Monsey, NY: Criminal Justice Press.

Ekblom, P. (1997). Gearing up against crime: A dynamic framework to help designers keep up with adaptive criminal in a changing world. *International Journal of Risk, Security and Crime Prevention*, *2*, 249–265.

Ekblom, P. (2001). *The conjunction of criminal opportunity: A framework for crime reduction toolkits*. London: Home Office. Retrieved on May 4, 2002, from www.crimereduction.go.uk/learningzone/cco.htm

Ekblom, P., Law, H., & Sutton, M. (1996). *Safer cities and domestic burglary*. Home Office Research Study #164. London: Home Office.

Field, S. (1993). Crime prevention and the costs of auto theft: An economic analysis. *Crime Prevention Studies*, *1*, 69–92.

Hope, T. (1995). Community crime prevention. In M. Tonry & D. P. Farrington (Eds.), *Building a safer society: Strategic approaches to crime prevention*. Crime and Justice, A Review of Research, Vol. 19. Chicago: University of Chicago Press.

Hesseling, R. (1994). Displacement: A review of the empirical literature. *Crime Prevention Studies, 3*, 197–230.

ICPC. (1997). *Crime prevention digest*. Montreal: International Centre for the Prevention of Crime (ICPC).

Keoleian, G. A., & Menerey, D. (1993). *Lifecycle design guidance manual*. Ann Arbor, MI: National Pollution Prevention Centre, University of Michigan.

Knox, J., Pemberton, A., & Wiles, P. (2000). *Partnerships in community safety: An evaluation of phase 2 of the safer cities programme*. London: Department of Environment, Transport & the Regions.

Laycock, G., & Tilley, N. (1995). Implementing crime prevention. In M. Tonry & D. P. Farrington (Eds.), *Building a safer society: Strategic approaches to crime prevention*. Crime and Justice: A Review of Research, Vol. 19. Chicago: Chicago University Press.

Learmount, S., Press, M., & Cooper, R. (2000). *Design against crime: a report to the Design Council, the Home Office and the Department of Trade and Industry*. Cambridge: Cambridge University Publications Centre.

Levi, M., & Handley, J. (1998). *Prevention of plastic card fraud*. Home Office Research and Statistics Directorate Research, Findings No.71. London: Home Office.

Maguire, M. (1997). Crime statistics, patterns, and trends: Changing perceptions and their implications. In R. Morgan & R. Reiner (Eds.), *The Oxford handbook of criminology*. Oxford: Clarendon Press.

Manunta, G. (1998). *Security: An introduction*. Cranfield, UK: Cranfield University Press.

Pease, K. (1998). Changing the context of crime prevention. In P. Goldblatt & C. Lewis (Eds.), *Reducing offending: An assessment of research evidence on ways of dealing with offending behaviour*. Home Office Research Study No. 187. London: Home Office.

Southall, D., & Ekblom, P. (1986). *Designing for car security: Towards a crime free car*. Crime Prevention Unit, Paper #40. London: Home Office.

Tonry, M., & Farrington, D. P. (Eds.). (1995). *Building a safer society: Strategic approaches to crime prevention*. Crime and Justice: A Review of Research, Vol. 19. Chicago: University of Chicago Press.

Welsh, B. C., & Farrington, D. P. (1999). Value for money: A review of the costs and benefits of situational crime prevention. *The British Journal of Criminology, 39*(3), 345–368.

Breaking the Cycle: Fundamentals of Crime-Proofing Design

by

Rachel Cooper
The University of Salford, The Adelphi Research Institute for Creative Arts and Sciences

Andrew B. Wootton
University of Salford, Design and Innovation Group

Caroline L. Davey
University of Salford, Design and Innovation Group

Mike Press
Robert Gordon University, Gray's School of Art

Abstract: *"Design Against Crime" aims to embed crime prevention within design through education and professional practice, and reflects the widening agenda for design professionals. This paper considers design as an essential contributor to product experiences, including that related to crime and fear*

Crime Prevention Studies, volume 18, pp. 179–201.

of crime. This paper argues that crime issues must be considered within design and new product development processes. Four fundamental principles of design against crime are presented: consult, develop, test and deliver. As part of the development and testing stages, ideas and concepts need to be generated that address causes of crime, as identified by criminologists. This paper presents the Design Policy Partnership's Crime Life-Cycle Model to help designers understand and effectively address causal mechanisms. The authors conclude that further design-centred resources are required to promote a more empathetic and holistic approach to crime prevention.

INTRODUCTION

Our material world is created through a design process, frequently through the employment of designers. Such designers employ their professional skills to understand the needs of the consumer or the user, and the technologies available, and then to develop a product, place, system or service which satisfies consumer needs and desires (Press & Cooper, 2003). Crime certainly occurs within this environment. Indeed, there are some who argue that designers are contributors to crime in that they develop desirable products, or products that can be used to commit a crime (Design Council, 2003). It seems that designers are the best placed to address crime effectively, and have the skills to do so. Pease (2001, p. 27), observes that:

> Designers are trained to anticipate many things: the needs and desires of users, environmental impacts, ergonomics and so on. It is they who are best placed to anticipate the crime consequences of products and services, and to gain the upper hand in the technological race against crime.

Among the design profession and manufacturers in many industrial sectors, crime prevention is often not considered until after a crime has occurred, rather than during the design stage of the development project (Learmount et al., 2000). This frequently leads to crime prevention solutions being "bolted on," after production, rather than embedded within the design. This is inconvenient, and neither an aesthetic solution nor a cost effective way of preventing crime (Town et al., 2003). In addition, once a product becomes subject to crime, it potentially increases fear of crime and thus consumers' decision to buy.

"Design-centred solutions" to crime are those approaches which centre around the design and development of the products. The designers are using their skills and knowledge of the crime and the situation to develop "products" with "inherent" crime prevention aspects. Design-centred solutions are more focused on the role of human behaviour, attitudes and emotions in preventing crime and feelings of insecurity. Solutions can be tailored to their specific context and address crime problems in innovative, often subtle ways. The involvement of design professionals enables a more empathetic and holistic approach that considers not only the potential misuse and abuse of products and environments, but aesthetics and human sensory experience. For example, a subtle solution to young people congregating and "hanging out" around public amenities—a frequent cause of anxiety, especially for older people—has been the use of classical music played softly in problem areas. Unlike other solutions, such as fencing or removal of public seating, it does not harm the visual aspect, nor inconvenience legitimate users. Another more pro-active solution is the development of a "youth shelter" that provides a place where young people can gather safely, without causing distress to local residents (Town et al., 2003).

In product development, it is critical to ensure that all requirements of the product's use, management and maintenance have been considered at the "front-end," for it is here that 80% of all subsequent costs are committed (Farish, 1992). Clearly this infers that crime should be considered early in the design process as a potentially negative factor that may impact on the user's total *design experience*.

THE DESIGN EXPERIENCE

The approach of design professionals to the activity of designing has broadened over the years, embracing concepts such as technology-led design, marketing-led design and consumer-led product development. Thus, we have "user-centred design," where extensive use of ethnography, anthropology and social science research methods is made to understand the user and inform the design process.

In addition to this widening design philosophy, there has been a move away from the product as the focus of attention for designers and their clients, to designing and delivering for the total experience (Press & Cooper, 2003).

The cycle of experience (see figure 1) starts at *life context*—the cultural and social background to any new design. This includes people's behaviour, patterns of living and working, shared cultures, concerns and beliefs, together with all the other products and innovations that help to form that

Figure 1: The Design Experience Model.

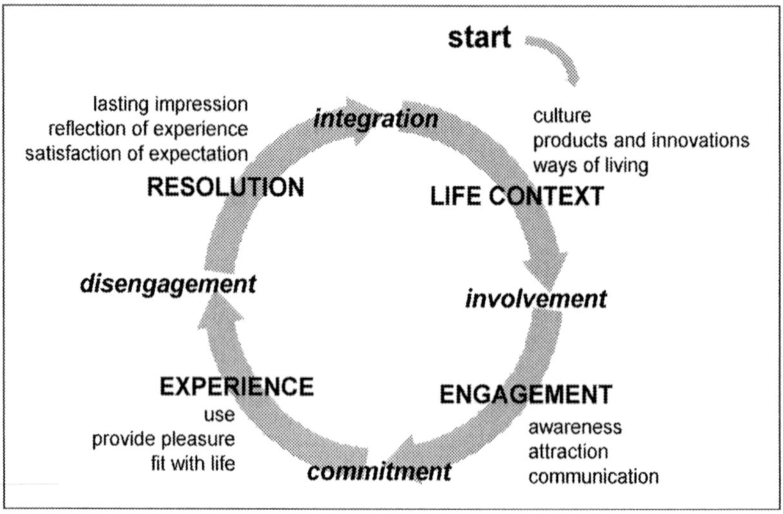

Adapted from: Rhea, D. (1992). A new perspective on design: Focusing on customer experience. *Design Management Journal, 9*(4), 10–16. Reprinted in M. Press & R. Cooper (2003), *The design experience.* Ashgate: Aldershot.

context. Our life context helps determine our responses to any design, and the different ways in which we may use or experience it. The central importance of life context explains why global companies increasingly commission extensive cultural lifestyle studies to ensure their designs adapt to changing values and meet the needs of emerging patterns of behaviour. In the *engagement stage*, the product must make people aware of its distinctive presence and attract people to it. The *experience stage* is where the product is used and becomes part of the user's life experience. This experience relates not just to the product's functionality and aesthetics, but also to the brand values and through them our own self-image. The user's design experience will embrace the entire product system and service—from the purchasing experience of the store, through to product ownership and service support. The *resolution stage* deals with the user's decision to "disengage" from the product, perhaps by upgrading, replacing or even recycling (Press & Cooper, 2003).

Crime and fear of crime is generally a negative factor in a user's design experience. As well as affecting life context, attraction to a product and attitude to purchase may be affected by perceptions of crime. The experience of the environment in which we purchase the product, whether that be retail or virtual, and the risk of crime when using the product, will colour our attitude toward both the product and the company.

THE WIDENING DESIGN AGENDA

Designers are increasingly being asked to consider issues outside of their traditional remit. This is moving them from dealing with merely economic and aesthetic issues and on to addressing complex social problems, including environmental and ecological issues, inequality and social exclusion. In response, specific design approaches have emerged, such as eco-design (Simon, 1999). This focuses the designer on the environmental impact of the materials used in the design, its manufacture and use. Indeed, current social and political pressures are encouraging designers to consider social responsibility and sustainability issues (Collings, 2003).

Crime and fear of crime is an area that designers must begin to address as a matter of course, and steps have been taken to raise awareness and support designers in designing against crime. In 1999, a national programme of research and policy initiatives emerged, with the aim of embedding crime prevention within design through education and professional practice. The programme was entitled *Design Against Crime* (DAC). Initiated by the U.K. Home Office, Design Council and Department of Trade & Industry, a major investigation of best practice was conducted (Learmount et al., 2000) and new design concepts developed. Institutions involved included the University of Cambridge, Central St. Martins College of Art & Design and the Design Policy Partnership—a multi-disciplinary team of researchers at the University of Salford and Sheffield Hallam University.

The Design Policy Partnership conducted research into design-led crime prevention, developing case studies (Cooper et al., 2002; Davey et al., 2002) and professional development materials. The Partnership also ran a national design competition for student designers and developed educational materials for school children that integrate with the National Curriculum (www.designagainstcrime.org). With funding from the European Union (EU) Commission (Hippokrates programmes 2001 and 2002 and AGIS programme 2003),[1] the Design Policy Partnership has conducted wider research in Europe and established an international network of partner organisations. This paper uses a synthesis of the results of this work.

The focus of the Design Against Crime program is "design thinking," the design process and design practice—in all its forms. Design Against Crime draws attention to the role that product, communication, interior, environmental and even fashion design can have in crime prevention. It highlights some of the more subtle and innovative ways in which designers can tackle crime and fear of crime. Consequently, Design Against Crime extends the scope of Crime Prevention through Environmental Design (CTPED) and national Secured by Design schemes into the wider arena of design practice, helping to bring design professionals from a range of disciplines into contact with established crime prevention networks, such as the U.K. and European Designing Out Crime Associations (DOCA-UK and E-DOCA).

For designers, crime is just one issue amongst many that must be addressed during a project. For Design Against Crime to be effectively integrated within design practice, its interface with the design process must therefore be understood.

THE DESIGN PROCESS

The design process has been extensively documented (Cooper & Press, 1995), as have models of the way in which designers think (Lawson, 1990). Generally, the design process entails four steps: Brief, Design, Test, Produce (Walker et al., 1989), which commonly sit within the new product development (NPD) process. This is described by Cooper (1994), as a stage-gate, structured process and latterly as a more fluid system (see figure 2).

Each stage in the process includes activities undertaken by a number of organisational disciplines such as marketing, production and finance. NPD theory proposes that designers work with the marketing function during the research and launch stages, and with production during the manufacturing stage. This increases the designer's understanding of market and user needs and technology and manufacturing processes available. The designer can therefore be considered central to decision making in terms of the overall product concept and its realisation. Clearly, this places the designer in the best position to address issues of crime prevention within the NPD process. To do this, the designer must have a grasp of crime issues and criminal causality. The Design Policy Partnership has developed a Life-Cycle Model to help designers understand criminal causality and develop solutions. However, for maximum effectiveness, Design Against

Figure 2: Stage-gate NPD Process Model.

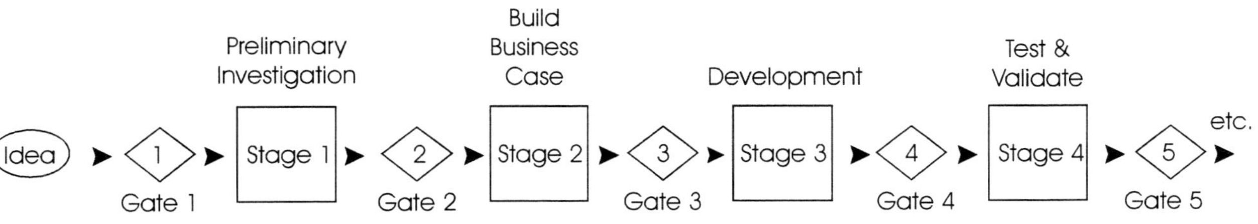

Today's stage-gate process.

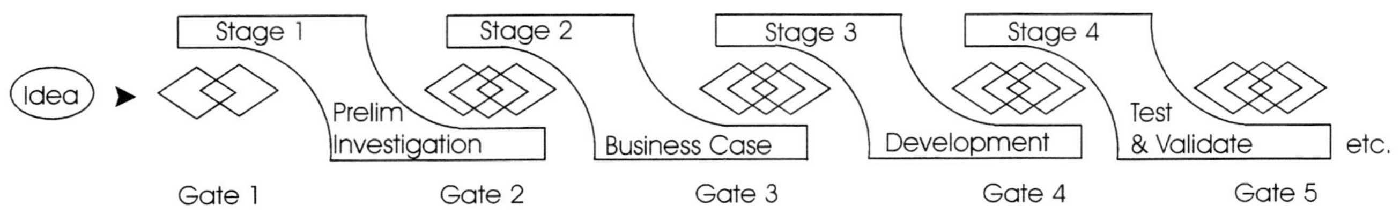

Tomorrow's Third Generation Process, with overlapping, fluid stages and "fuzzy" or conditional Go decisions at gates.

Cooper, R. G. (1994). Third generation new product processes. *Journal of Product Innovation Management*, *11*, 3–14.

Crime should be integrated within the product development process. The following section outlines the *four fundamentals* to designing against crime (Cooper et al., 2003; Design Council, 2003).

THE FOUR FUNDAMENTALS OF DESIGN AGAINST CRIME

From analysis of extensive case study research, four fundamental steps have been identified that enable design against crime thinking to be embedded within the New Product Development process (Design Council, 2003; Cooper et al., 2003). These four steps are briefly outlined below, and are described in more detail in the Design Council's (2003) publication "*Think Thief: A Designer's Guide to Designing Out Crime.*"

Stage 1: Consult

There is a growing awareness within organisations of the necessity to adopt a more responsible approach to the way in which they develop, manufacture and market products. However, there are still those in industry who are uncertain as to how they can effect change both within their organisation and its wider social environment. Organisations will not be fully committed to Design Against Crime unless they can see the benefits. This stage of Design Against Crime involves winning support for, and gaining input into, the design process from people both inside and outside the organisation. Design Against Crime strategies can reward organisations in a number of ways. For example:

- Enabling competitive advantage
- Protecting their brands
- Cutting the cost of theft
- Reducing crime against employees and customers
- Supporting local communities.

Designers, other team members and senior management need to be committed to the process of Design Against Crime. When the needs and requirements of stakeholders are fully understood, the opportunities for Design Against Crime can be identified and communicated.

In developing a Design Against Crime strategy, the following questions may help identify the constraints and opportunities.

- What are the emerging crime trends related to your products or services?
- Have there been any technological developments that could make your products or services more vulnerable or secure?
- Do your customers have any suggestions as to how you might crime-proof your products or services?
- Have you gathered feedback regarding crime and fear of crime from other information sources, such as maintenance staff and insurance companies?
- Do you undertake crime risk assessments as part of the NPD process?
- Do you gather information about crime?

There are a wide variety of stakeholders who can provide crime-related information that informs NPD strategy and practice. Stakeholders should be identified at an early stage. They will include internal stakeholders, such as research, design, marketing and sales staff, and external stakeholders, such as customers, end users, suppliers, the police, crime prevention agencies and statutory agencies. It is particularly important to gain an understanding of offenders' motivations and *modus operandi*. Those dealing with offenders, such as police and probation officers, will be particularly useful in this respect.

Stage 2: Develop

Design Against Crime solutions must be tailored for their specific context. This phase of Design Against Crime involves research to gain an in-depth understanding of potential crime risks and problems associated with the product, environment or service being designed. The risks of different crimes occurring need to be assessed, including theft, burglary, criminal damage, violence, fraud, forgery and robbery. Information should also be gathered regarding the proposed design solution in relation to the environment where crimes might occur and the characteristics, behaviours and motivations of potential offenders.

Designers must consider the potential for their designs to be misused or abused. So they need to think not only about the *user*, but also about the *abuser* or *misuser*. To achieve this, designers need to learn to *think thief*—to anticipate potential crimes and offenders' actions, understand their tools, knowledge and skills and incorporate attack testing into the design process. This means not just thinking thief, but considering all types of crime (Town et al., 2003). Offenders typically ask a number of key questions, when preparing to commit a crime: Can I be seen? If I am seen, will I be noticed? If I am seen and noticed, will anybody take any action? Do I have an easy escape route?

The design under development may become a target for criminals. This will largely depend on its design attributes. For example, in the case of theft, a product is more likely to be stolen if it is CRAVED; that is Concealable, Removable, Available, Valuable, Enjoyable and Disposable (Clarke, 1999). It is particularly important that crime issues are considered in relation to high-risk products and environments.

The design of products, places and messages impact on actual crime and the fear of crime by influencing the attitudes and behaviours of:

- Potential offenders
- Formal guardians, such as police officers and private security personnel
- Informal guardians
- Potential victims and targets of crime
- Victims of fear of crime.

Information is available that details good Design Against Crime practice, such as guidelines (Town et al., 2003; ODPM, 2004) and case studies (Davey et al., 2002). Nevertheless, solutions applied elsewhere, however successfully, will need to be tailored to the context and requirements of the current situation. Designers should also think about the potential countermeasures offenders may adopt as a result of a design intervention. The aim is to *out-think* the offender and develop design solutions that "short-circuit" the potential offender's behaviour, but without reducing the design's value to legitimate users, increasing fear of crime, creating social problems, or causing the seriousness of the crime to escalate (Town et al., 2003).

The Design Policy Partnership have developed a Crime Lifecycle Model to assist designers in developing concepts and ideas that address the causal factors that underpin crime, and also to think about issues that may occur after a crime has been committed. The Crime Lifecycle Model also enables designers to test their design solutions against criminological ideas of causality (Ekblom & Tilley, 2000). This is described in more detail later in the next section.

Stage 3: Test

All design solutions must be fully validated in terms of addressing appropriate crime issues and "short-circuiting" specific causal factors before they are made available for use. Traditional techniques for product assessment, such as SWOT (Strengths, Weaknesses, Opportunities, Threats) and cost-benefit analyses, should be given a crime perspective. User trials and performance testing need to encompass both use and potential misuse of the design solution. Focus groups to test fitness for purpose might include crime prevention experts, such as police architectural liaison officers, crime prevention design advisors, criminologists, community representatives, potential offenders and ex-offenders. In addition, designers should compare their solutions to existing good practice, and consider "what if?" future scenarios.

Stage 4: Deliver

Design Against Crime offers tangible commercial benefits—crime resistance can be a good marketing tool. Crime-resistant design features offer a "double win," in that they can not only reduce the incidence of crime, but also provide users with a sense of security, reassurance and peace of mind that positively impacts on the design and brand. Vehicles for promoting Design Against Crime solutions include specialist magazines and journals, national and local newspapers, television, radio and participation in conferences and exhibitions. Accreditation schemes exist for organisations and designers wishing to formally demonstrate their commitment to addressing crime (e.g., Secured by Design, in environmental design sector).

In addition, building in resistance to crime can reduce maintenance costs and increase the longevity of a design. The response of users and the public to crime-related issues can vary, however, and this must be carefully assessed when developing any delivery and marketing strategies.

The impact of crime on a design solution should be evaluated over its life, and changes in the type of criminal activity perpetrated against it monitored. Unforeseen use of the product for criminal ends may emerge once it is in the market. Such information should be fed back to the organisation and designers to inform design decisions relating to later versions of the product, environment or service design.

The next section of this paper will describe in more detail the approach to design-led crime prevention developed by the Design Policy Partnership to help designers understand and address the causes of crime.

MODELS OF CRIMINAL CAUSALITY

Good design reduces criminal opportunity—the main cause of crime. Denying opportunities to potential offenders increases a crime's difficulty, and decreases its attractiveness and potential reward (Felson & Clarke, 1998). Criminologists have written a great deal about the causes of crime. In general, *causal factors* for particular crimes have been offender-focused and can be divided into two main areas:

- those factors relating to the attributes of the offender who commits the crime; and,
- those factors relating to the attributes of the situation in which the crime happens.

The aim of categorising the causal factors that underpin crime is to enable criminologists to better understand the influences and choices affecting offender behaviour in a crime situation. One such causal framework, developed by Ekblom (2001) at the U.K. Home Office, has been simplified and extended for use by designers.

Active versus Passive Design

When crime is not considered during the design process, the designed product becomes passively subject to offender-related attributes and situation-related attributes that lead to crime (see figure 3).

When designers consider crime effectively, aspects of the product being developed are designed to counter or *block* the offender- and situation-related attributes that lead to crime (see figure 4).

Of course, not all crime can be prevented by design. Criminals are becoming ever more creative in the countermeasures they employ against crime prevention measures. So given that some crime will "get through" any design-led prevention, it's sensible for designers to embody measures within their designs that consider issues beyond the crime event, including: helping identify, apprehend and prosecute offenders and mitigating against the short- and long-term effects of crime—both for the victim and the offender. This paper will go on to discuss a crime lifecycle model and design process that has been developed for designers and product developers to consider the wider issues related to design against crime.

Figure 3: Passive Design.

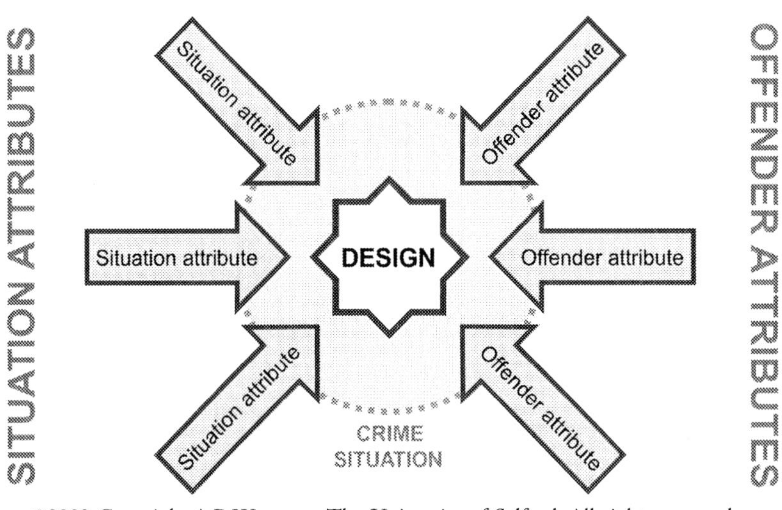

THE CRIME LIFECYCLE MODEL

Based on the thinking outlined above, a *lifecycle* model for crime has been developed (Wootton et al., 2003) to help design professionals address crime issues during the development of design concepts (see figure 5).

The Crime Lifecycle Model embodies three key principles:

1. *Offending behaviour can breed further offending behaviour, so the real key to sustainable crime prevention is to break this cycle*—The initial and final phases of the model are linked, as basic "readiness to offend" is fuelled largely by a potential offender's life circumstances and the extent to which they believe criminal behaviour to be acceptable and a valid option for them. The longer-term consequences of criminal activity will affect an offender's readiness to offend. For example, the difficulties of finding employment experienced by those with a criminal record may encourage further criminal behaviour. In addition, being locked up with a large number of other criminals may provide an education in criminal skills, and access to resources for future activities.

Figure 4: Active Design.

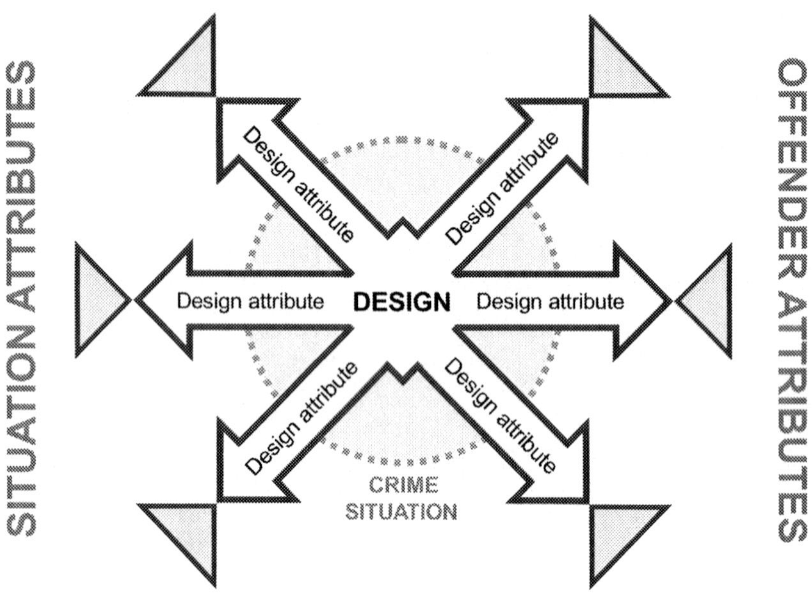

2. *All six pre-crime issues (Phases 1 to 6) are prerequisite to a crime event occurring*—By comprehensively addressing any one of these issues, the crime event can be prevented from occurring.

3. *Post-crime issues should also be considered*—As already mentioned, it's unlikely that any measure will be 100% effective. Therefore it is important to consider how the product, service or environment might be designed to address issues that arise after the crime event (Phases 7 to 10).

Phase 1: Offender's readiness to offend–This relates to aspects of a potential offender's life circumstances, including his or her:

- Financial situation
- Employment situation
- Family circumstances

Figure 5: The Crime Lifecycle Model.

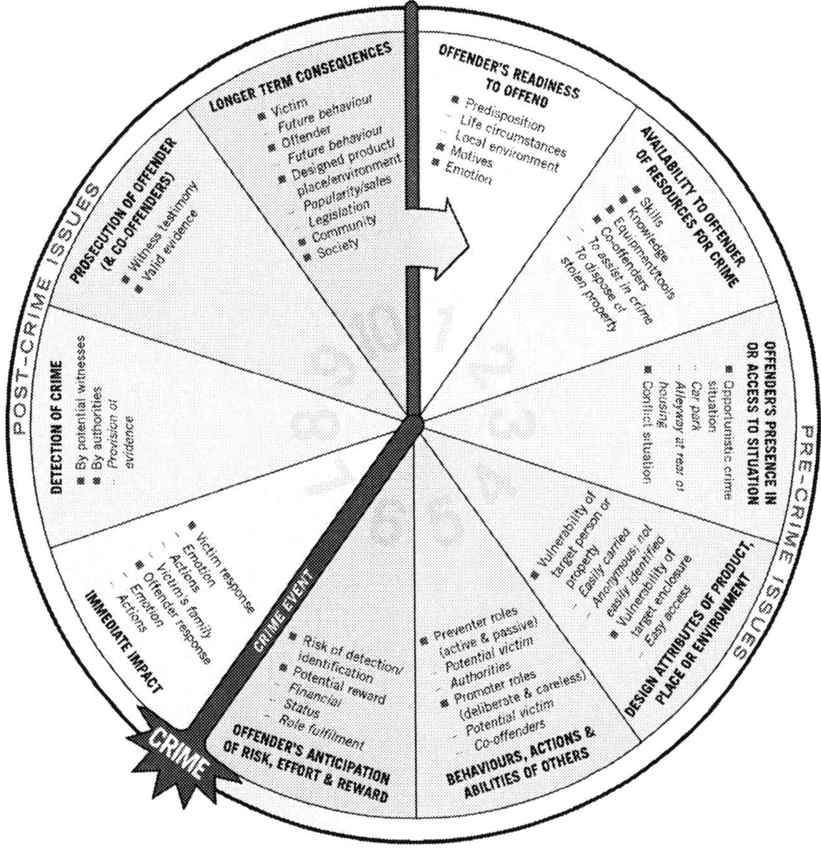

- Personal relationships
- Education
- Beliefs
- Motivations.

One example of how design can tackle this issue is through the regeneration and redevelopment of poorly designed areas. An example of this is the Royds Community Association case study (Davey et al., 2002), where not only were local residents involved in the planning and design process, but local people were employed on its construction. Such involvement significantly improves feelings of community, shared ownership and responsibility for the built environment. This residential area in the North of England suffered high levels of unemployment and alcoholism, as well as drug addiction and burglary. Innovative methods had to be used to engage local residents in the consultation process, as many residents were too fearful of crime to attend public meetings. (Figure 6 shows a meeting between designers and planners and local people). Engaging the users in the design can be undertaken with most products and services.

In addition, this pre-crime issue includes negative emotions such as anger and hate. Such emotions can motivate particular offending behaviour.

Phase 2: Availability to offender of resources for crime–This issue relates to the resources that an offender will employ to commit crime. Such resources include physical items such as tools and equipment, but often the main resource in the offender's knowledge, skills and abilities to commit a particular offence. The more experienced the offender, the more skilled he or she becomes.

Co-offenders may also be considered a resource in so much as they:

- assist before or at the time the offence is committed (for example, assisting in a burglary); and/or,

Figure 6: Royds Community Association Involvement in Regeneration.

Reprinted with permission of the Royds Community Association.

- assist after the offence has taken place (for example, buying stolen goods).

Designers should be aware that offenders may use their designed product to resource their criminal behaviour—and should design against this. An example of this is given in the AlphaBar case study (Davey et al., 2002), where the use of toughened glass prevents the use of beer glasses as weapons in bar fights. Toughened glass shatters into many hundreds of tiny pellets when broken. Conventional glass, by contrast, serves as an offensive weapon because it leaves larges, potentially lethal pointed edges.

Phase 3: Offender's presence in or access to situation–The focus of phase is the presence—purposeful or opportunistic—of a potential offender in a crime situation or conflict situation. Regarding offenders presence in conflict situations, the Wendover public house case study (Davey et al., 2002) gives an example of pub interior that was redesigned to reduce the incidence of arguments that led to crimes of violence. Bar stools were removed from the bar area to decrease crowding and reduce the likelihood of drinks being spilt—a frequent cause of fights breaking out in pubs, bars and clubs. The toilet entrance was relocated further into the pub lounges, away from the main entrance, where bar staff could better observe customers. This meant that the presence of bar staff deterred those engaged in drug dealing from using the pub toilets for their transactions.

Phase 4: Design attributes of product, place or environment–The issues in this phase relate directly to the vulnerability of the designed product, place or environment to crime. In the case of product design, this includes whether the product is CRAVED (concealable, removable, available, valuable, enjoyable and disposable).

Perhaps the most vivid examples of designing against crime in this phase are the various measures taken by automotive manufacturers in recent years. According to recorded crime figures, car crime increased steadily from 1979 to a peak in 1992, and has since been reducing at a reasonably steady rate. Currently, the figures are at their lowest levels since 1989. Much of this decline has been attributed to the increased security now being designed into new vehicles, such as alarms, immobilizers and tracking systems (Learmount et al., 2000).

Phase 5: Behaviours, actions and abilities of others–This phase considers the actions of others in a crime situation—either as preventers or promoters. Preventers can be either active (for example, a resident who challenges a stranger acting suspiciously in their neighbourhood) or passive (for example, the mere presence of other people may deter an offender from mugging someone). Similarly, crime promoters can be either deliberate (for example, the accomplice who keeps watch while the offender burgles a house) or careless (for example, the person who forgets to lock their car, or leaves their mobile phone on the desk in the office).

Communications design can challenge attitudes, empower victims and generate support for those targeted by offenders. A poster campaign was developed to raise awareness of domestic violence—a much under-reported crime affecting one in four women from all social backgrounds. This was undertaken by Creative Medialab—a partnership between Creative Input Limited and Buckinghamshire Chilterns University College—working to a brief from Wycombe Women's Aid. The posters created are both hard-hitting and provocative, and primarily aim to increase crime reporting by victims and witnesses.

Phase 6: Offender's anticipation of risk, effort and reward–Phase 6 concerns the offenders' immediate motivations for committing an offence, including:

- their perception of the risk that they may be detected;
- their perception of the risk that they may be identified; and,
- their perception of the potential reward the offence offers. This may be financial, but may equally well be about status within their peer group.

Phase 7: Immediate impact–This phase is concerned with the immediate consequences of the crime event. Designers should think about the potential impact on both the victim and the offender, how they may react, and if negative reactions can be modified in any way by design. It is here that potential negative results of design interventions during the pre-crime phases should be considered, and steps taken to ensure that impact of crime for the victim is not inadvertently made worse.

Phase 8: Detection of crime–Phase 8 is concerned with the detection of crime by witnesses and ultimately by the police. Such detection should provide adequate evidence to enable prosecution of the offender. Examples of how this issue can be addressed include the use of CCTV, and marking property with the owner's postcode using ink that is visible under ultra-violet light.

Phase 9: Prosecution of offender (and co-offenders)–This phase is concerned with the successful prosecution of the offender and any co-offenders. To address this issue, designers should think about how their designs might support the provision of witness testimony and valid evidence. Examples of design interventions in this phase include the use of fingerprint sensitive paints and permanent dyes and markers to link offenders to crime scenes. SmartWater is a forensic coding system, which can be applied to valuable possessions, safely and discreetly. The unique formula allows processions to be forensically identified. Forensic profiling can directly link offenders to a crime (see website: http://www.smartwater.com/news/fom.html).

Phase 10: Longer-term consequences–This phase is concerned with alleviating the longer-term consequences of crime such as increased fear on the part of victims and their families. Participation in crime prevention can empower individuals and communities, as well challenge the legitimacy of offending behaviour. New legislation is placing greater onus on Local Authorities for crime prevention.

CONCLUSION

This paper showed that designers have a key role to play in crime prevention. They have the skills to both anticipate crime risks (Pease, 2001), and develop innovative solutions to crime. Furthermore, the design process focuses on identifying and fulfilling user needs and requirements through an understanding of human behaviour, attitudes and emotions in relation to a particular design objective, enabling designers to adopt a "human-centred" approach to preventing crime and feelings of insecurity. Good design solutions are tailored to their specific context and often address crime problems in subtle, creative ways. The involvement of design professionals supports more effective integration of crime prevention within design solutions, enabling a move away from the simple retro-fitting of security devices (Town et al., 2003).

With growing interest in sustainability and social responsibility and their focus on the user's total experience, it is timely for designers to consider crime as an integral part of the new product development process.

Design Against Crime encourages an empathetic, human-centred and holistic approach that considers not only the potential misuse and abuse of products and environments, but aesthetics and overall sensory "design experience."

Over the last decade there has been a shift towards "cyclical thinking" in both design and crime prevention. The "Design Experience" Model developed by Press and Cooper (2003) reflects this approach. In terms of the former, the shift has been driven by the need to consider how design influences all aspects of the consumer experience, from life context and lifestyle, through brand identification, usability, and brand allegiance. In terms of the latter, the crime lifecycle has developed as a useful model to consider the inter-relationships between interventions and actions targeted at the causes, conditions and consequences of crime events (Wootton et al., 2003). It is an extremely useful framework for considering how design thinking can focus at the various dimensions of crime.

We have also sought to develop our thinking, and the outcomes of the research, from an emphasis on process to that of strategy. Over the course of our research there has been a significant increase in awareness of the role of design and designers, and through the research initiative we now have tools and techniques to apply within the design process. Yet "crime proofing" will not be embedded in the product development process until all the agencies understand the significance of using these concepts and the designers have the knowledge and skills to do it.

This paper has outlined the fundamental stages of designing against crime: consult, develop, test and deliver. These stages of Design Against Crime can be easily superimposed onto those of the wider new product development process (Cooper, 1994; Farish, 1992), making the adoption of Design Against Crime by design professionals relatively uncomplicated. It is shown that some activities within the design process, such as focus group research and cost-benefit analyses, need to address crime issues and to involve crime prevention experts. For this to occur, designers will require access to appropriate data sources and knowledgeable contacts.

When it comes to developing design solutions that address the causes of crime, however, designers need access to design-centred resources that draw more deeply on criminological theory and practice. As an example, this paper presented the Crime-Life Cycle Model, based on Ekblom's (2001) work on the causes of crime. The authors believe that further resources are required to help designers address crime effectively during the design process.

Finally, safety is a key quality of life issue, and must be considered alongside other sustainability issues such as environmental quality, economic diversity and social inclusion. The next step must be to bringing together multiple disciplines to take a holistic perspective to bring science, technology, social science and design together in pursuing research which includes safety as part of the sustainable design agenda.

Acknowledgments: The Design Policy Partnership would like to thank its funders: European Commission AGIS 2003 Programme, Design Council, Home Office, Department of Trade & Industry and the Engineering and Physical Sciences Research Council (EPSRC).

This paper reflects the views of the authors. The funding bodies are not liable for any use that may be made of the information contained here in.

NOTE: To support the furtherance of this research agenda, the University of Salford is leading a Sustainable Urban Environments consortium—comprising University College London, Sheffield Hallam University and London Metropolitan University—in a groundbreaking research project entitled *VivaCity2020*. The aim of this five-year, EPSRC-funded project is to support and enable sustainable and socially responsible urban design through the development of innovative, inclusive and practical decision-making tools and resources. These will be derived from an in-depth understanding of the patterns of human/environment interaction, and will resolve practical urban design, operation and management problems, particularly in relation to the twenty-four hour city. VivaCity2020 will bring together multiple disciplines to address issues of social responsibility and sustainability, including science, technology, social science and design.

NOTE

[1]Hippokrates is a European Union funding programme intended to provide support for the European crime prevention strategy by encouraging cooperation among interested public and private-sector organizations. AGIS is a framework programme to help police, the judiciary and professionals from the EU member states and candidate countries cooperate in criminal matters and in the fight against crime.

REFERENCES

Cain, J. (1998). Experience-based design: Towards a science of artful business innovation. *Design Management Journal*, *9*(4), 10–16.

Clarke, R. V. (1999). *Hot products: Understanding, anticipating and reducing demand for stolen goods.* Police Research Series Paper 112. London, UK: Home Office Research, Development and Statistics Directorate.

Collings, R. (2003). *About: Corporate Social Responsibility.* Downloadable PDF document available from: www.designcouncil.org.uk

Comite Europeen de Normalisation (CEN). (2002). *A European standard for the reduction of crime and fear of crime by urban planning and building design: ENV 14383-2.* Brussels, Belgium: CEN (www.cenorm.be/aboutcen.htm).

Cooper, R. G. (1994). Third generation new product processes. *Journal of Product Innovation Management, 11*, 3–14.

Cooper, R., Davey, C., & Press, M. (2002). Design against Crime: Methods and issues that link product innovation to social policy. *International Journal of New Product Development & Innovation Management, 3*(4), 329–342.

Cooper, R., Davey, C. L., Wootton, A. B., Heeley, J., Press, M., & Hands, D. (2003). *Design against Crime: Guidance for professional designers.* Report to the Design Council. Salford, UK: The University of Salford.

Cooper R., & Press, M. (1995). *The design agenda* (pp. 36–43). Chichester: John Wiley and Sons.

Davey, C. L., Cooper, R., & Press, M. (2002). *Design Against Crime case studies.* Salford, UK: The Design Policy Partnership (University of Salford and Sheffield Hallam University).

Design Council. (2003). *Think thief: A designers guide to designing out crime.* London, UK: Design Council.

Ekblom, P. (2001). *Crime reduction: The conjunction of criminal opportunity.* London: Home Office Crime Reduction Toolkits (www.crimereduction.gov.uk/cco.htm).

Ekblom, P., & Tilley, N. (2000). Going equipped. *British Journal of Criminology, 40*, 376–398.

Farish, M. (Ed.). (1992). *New product development: The route to improved performance* (p. 2). London, UK: Design Council.

Felson, M., & Clarke, R. V. (1998). *Opportunity makes the thief: Practical theory for crime prevention.* Police Research Series Paper 98. London, UK: Home Office Research, Development and Statistics Directorate.

Lawson, B. (1990). *How designers think.* Oxford, UK: Butterworth-Heinemann.

Learmount, S., Press, M., & Cooper, R. (2000). "Design Against Crime." Report to the Design Council, Home Office and Department of Trade and Industry.

ODPM (Office of the Deputy Prime Minister). (2004). *Safer places. The planning system and crime prevention.* London, UK: Thomas Telford.

Pease, K. (2001). *Cracking crime through design.* London, UK: Design Council.

Press, M., Cooper, R., Davey, C., Heeley, J., & Wootton, A. B. (2003, April). *The wrong trousers: A critical reflection on design and crime prevention.* Paper presented at the European Academy of Design conference, Barcelona.

Press M., & Cooper, R. (2003). *The design experience* (p. 95). Chichester, UK: John Wiley and Sons.

Rhea, D. (1992). A new perspective on design: Focusing on customer experience. *Design Management Journal, 9*(4), 10–16. Reprinted in M. Press & R. Cooper (Eds.), (2003), *The design experience.* Aldershot, UK: Ashgate.

Simon, M. (1999). *Ecodesign cultures in industry.* Paper presented at the third European Academy of Design conference, Sheffield Hallam University, 30 March–1 April 1999.

Town, S., Davey, C. L., & Wootton, A. B. (2003). *Secure urban environments by design: Guidance for the design of residential areas.* Salford, UK: The University of Salford.

Walker, D., Oakley, M., & Roy, R. (1989). *Managing design: Overview* (p. 34). (Teaching Material for Module P791.) Milton Keynes, UK: Open University Press.

Wootton, A. B., Davey, C., Cooper, R., & Press, M. (2003). *The crime lifecycle: Generating design against crime ideas.* Salford, UK: The University of Salford.

Designing out Crime from the U.K. Vehicle Licensing System

by

Gloria Laycock

and

Barry Webb

Jill Dando Institute of Crime Science, University College London

Abstract: *The U.K. has one of the most user-friendly vehicle registration and licensing systems in the world. As a system for ensuring compliance with vehicle taxation and other regulations, however, it has problems. It has evolved over many years, with changes being introduced in piecemeal fashion as vehicle regulations were revised or new ones instituted. The ad hoc way in which the system has developed has meant that it has become less effective than it might be as a taxation and control instrument. Many problems are systemic in that they are a consequence of the design of the system rather than resulting from any inefficiencies in its administration. This paper describes the findings of a review, commissioned by the British government, to examine how the system might be redesigned to make it fit for the 21st century. The review showed how system design changes could be introduced to ensure better compliance by users. This paper also examines the experience to date in trying to implement some of these changes.*

Crime Prevention Studies, volume 18, pp. 203–230.

INTRODUCTION

It is now fairly well understood that the design of goods can facilitate crime and increase offending by providing the opportunities that cause crime. It is perhaps less well appreciated that the design of management systems and of legislation can have a similar effect. For example, the way in which goods are displayed in self-service stores makes shop theft easier at the same time as increasing sales. The astute manager has to achieve a balance between making the goods appear attractive and available and making their theft that much easier, with the obvious consequence.

There are a few studies that illustrate this point. Levi and Handley (1998) have, for example, demonstrated that the design and use of credit cards has had a major effect on the extent and type of credit card fraud. Similarly, Burrows (1991) has shown how parts of the private sector have modified aspects of their operations in order to minimise the extent to which they are victims of crime, sometimes by even collaborating with competitors. More recently Newman and Clarke (2003) have considered the way in which the Internet has developed without too much initial thought of the crime-producing aspects of its operation. Finally, a recent report for the European Union (EU) (Jill Dando Institute, 2003a) has shown that poorly drafted legislation can cause crime problems as much as reduce them. For example, legislation that makes it hard to dispose of dangerous substances creates a market for their disposal which organised criminals have been quick to capitalise upon (Webb & Marshall, 2004).

In this chapter we describe the results of what was in effect a consultancy, commissioned in the U.K. by the then Department of the Environment, Transport and the Regions (DETR), now the Department for Transport (DfT) and referred to as such for the remainder of the chapter.[1] The study considered the U.K. vehicle licensing system in 2001 and made recommendations for its improvement. Crime reduction steps have tended to be introduced in a piecemeal way, rather than by considering which fundamental changes will address the deficiencies effectively. This study was intended to take a more holistic approach.

The chapter first describes the U.K. system at the start of the project and outlines the results of the consultancy, setting out the recommendations that were made. The way in which the Department dealt with these recommendations is considered in the final section of the chapter.

THE U.K. VEHICLE REGISTRATION SYSTEM IN 2000

The U.K. vehicle registration and licensing system has developed over time as a revenue-raising device, and for ensuring compliance with other vehicle regulations. Four important aims underlay the registration system in 2000 and still apply today. These are:

1. *Identification of motor vehicles and their keepers*–The registration system provides for the unique identification of motor vehicles through the use of registration marks and the Vehicle Identification Number (VIN). The Driving and Vehicle Licensing Authority (DVLA) vehicle record provides for the identification of keepers and links vehicle identification information with keeper details.

2. *Collection of vehicle excise duty*–The licensing system is the collection mechanism for vehicle excise duty. Payment is receipted through the issue of a tax disc to motorists. This tax disc is mounted on the inside of vehicle windscreens and demonstrates the payment of vehicle excise duty to law enforcement and other agencies. The vehicle license is not transferable between vehicles.

3. *Compliance check for compulsory MOT requirement*–Motorists seeking to renew their vehicle license must produce a current MOT certificate (called MOT because it was introduced by the then Ministry of Transport) for all vehicles over three years old. This certification is issued by an authorised agency following a comprehensive check on the roadworthiness of the vehicle. This is the only routine national enforcement of the MOT certificate.

4. *Compliance check for compulsory insurance requirement*–When a vehicle license is first taken out and at all vehicle re-licenses, motorists must produce valid insurance documents for inspection. This is the only routine check that the person using the vehicle is insured to drive. The insurance currently only has to be valid for the single day on which the insurance documents are inspected.

The U.K. system is one of the most user-friendly in the world and, perhaps as a consequence, it has not been as successful as it might have been as a taxation and control instrument. For example, it is estimated that in 2002 there was a 5.5% evasion rate for vehicle excise duty, with the associated loss of tax then estimated to be some £193 million per annum (Smith & Webb, 2005). It is also estimated that the incidence of driving without insurance in the U.K. is around 5% (Greenaway, 2004), which imposes a cost of around £30 per year on the premiums of honest motorists. This compares with an incidence of as low as 1% in some EU member states. Vehicle insurance fraud is an additional concern to insurers.

Abandoned cars are also becoming more of a problem. On average, some 1.8 million cars reach their end of life each year and are scrapped. However, it is estimated that, in 2000, 238,000 cars were abandoned on U.K. streets, leaving local authorities with an expensive removal problem. This is likely to increase significantly as European legislation on the disposal of vehicles is implemented (Smith et al., 2004). A reliable, up-to-date vehicle register, particularly of older cars, is an important component in dealing with this issue.

Finally, nearly 340,000 cars are stolen in the U.K. every year, of which about 120,000 are never recovered. Car crime accounts for a fifth of all recorded crimes and costs at least £3 billion annually. Many serious crimes including murder, rape, abduction and armed robbery rely on the ease with which vehicles can be stolen and remain outside the system, and the simplicity with which the apparent identity of the vehicles can be changed.

WHAT WE DID

The Jill Dando Institute of Crime Science (JDI) was asked to address these deficiencies, to consider which principles should underlie effective vehicle registration and licensing, and to make recommendations for a world-class U.K. system.

In order to obtain a balanced and thorough view of the system in 2001, we consulted a number of stakeholders or stakeholder organisations, which were initially identified through contacts provided by members of the project steering committee.[2] These stakeholders were contacted by telephone, email or in writing and were given details of the project and offered an opportunity to contribute their views either in writing, over the phone, or in person.

It was explained to stakeholders that the JDI was conducting an exploratory study into the vehicle licensing and registration system. They were asked for their views on the strengths and weaknesses of the system, particularly with respect to crime prevention; what modifications might be introduced to overcome the weaknesses; and what might be the main features of an ideal licensing and registration system.

The steering committee identified Sweden and The Netherlands as having innovative and successful licensing and registration systems. These countries were visited and reports summarising their operations were produced. We also consulted senior officials at the European Commission in Brussels. A consultancy company visited Austria to report on the system there, which was also felt to have some strengths.

Finally a specialist company was engaged to carry out a series of brainstorming workshops and focus groups. The company provided strategic advice and experienced facilitators for the workshops, having first met with the steering group to agree upon the topics. These were:

- How to identify vehicles
- How to minimise evasion of vehicle excise duty, MOT, and insurance and ensure proper vehicle disposal
- How to unambiguously link a car with a person
- How to ensure reforms do not lead to social exclusion.

Four brainstorming workshops were conducted with stakeholders and other interested parties covering each of the identified topics. The participants were identified through earlier contacts with stakeholders, through contacts provided by steering committee members, and through our own research. Individuals from a number of relevant organisations were contacted and asked if they would like to participate in the workshops or to suggest possible participants. It was explained to those contacted that creative individuals were required for the workshops and that individual input was being sought, not the views of their particular organisation. The objective of these sessions was to obtain creative solutions to the problems identified.

There were two focus groups, which were held with members of the public, to establish preliminary feedback on a small number of the potential solutions that had been identified from the list of ideas generated during the workshops. The selected material was presented to each of the groups by the group facilitator during the course of a semi-structured discussion of vehicle regulation.

Results

As a result of these combined exercises we identified a set of principles to which an ideal system should conform, and a set of next steps, which would move us toward such a system.

Principles of a Good System

The general principles listed in Table 1 apply to the whole system. We were particularly asked to avoid solutions that were dependent upon primary legislation. Whilst we accepted this constraint, it may not be entirely realistic if an "ideal" system is to be designed in the longer term. An appropriate legislative framework will form the foundation upon which the recommendations set out here will be based, and a willingness on the part of the government to implement that legislative change is therefore fundamental.

Table 1 Principles of a "Good" System

1. Can be implemented quickly
2. Minimal need for legislation
3. Cost effective
4. Future proof—responsive to changing circumstance
5. Default to honesty, i.e., more difficult to break the rules
6. Responsibility clear from "designer to shredder"
7. Privacy and data protection issues to be taken into account
8. Easy to use:
   One-stop shop
   Customer friendly
   No undue burden on partners in private sector
   Simple systems to ensure compliance and detect non-compliance
   Easy to enforce

Next Steps

The consultation process identified a number of problem areas from which we selected five that we felt could be addressed relatively quickly and that could make a significant difference to the problem:

- database inaccuracy;
- insecurity of vehicle identification systems;

- inadequate enforcement;
- lack of strategic overview; and,
- crime prevention as a priority.

RECOMMENDATIONS

Eleven specific recommendations were made in 2002 to address these problems. These are listed below under the relevant problem headings.

Problem 1: Database Inaccuracy

It is a fundamentally important characteristic of any good vehicle registration system that the vehicle register has accurate information concerning vehicles and keepers. If it worked properly, the register would be able to fulfil a number of functions to tackle different aspects of crime including:

- **Identifying vehicle excise duty evaders:** Accurate details of the vehicle keeper are required to follow up those who have not paid their vehicle excise duty.
- **Identifying stolen vehicles:** The police may identify stolen vehicles by checking the registration number with the vehicle record (via the Police National Computer) to see if there is a "stolen" marker on the database. In addition, in certain circumstances DVLA may become suspicious that a vehicle has been stolen and so refer the issue to the police. An accurate vehicle record is required to facilitate this process.
- **Identifying traffic offenders:** The police use the registration system to identify keepers of vehicles involved in traffic offences so that a fixed penalty notice or summons can be issued.
- **Identifying vehicles involved in crime:** The police may use details of registration numbers taken down by witnesses at the scene of a crime to follow up as lines of enquiry. The registration number will be linked to a vehicle that can be linked to a keeper.
- **Authenticating vehicle keepership:** The general public rely on the vehicle record as a means of authenticating that a vehicle is registered to the person selling it and that the vehicle being sold matches the official documentation held by the vendor.

- **Determining the keepers of abandoned vehicles:** Authorities use the vehicle record to check the keeper details of abandoned vehicles. Since some of these are vehicles involved in fraudulent insurance claims, this also helps detect insurance fraud.

The first problem was that the vehicle databases contained significant inaccuracies. The vehicle register may have contained over 10 million more vehicles than the number believed to be on the roads (Martell Consultants, 2002). The Motor Insurance Bureau (MIB) reported that their experience in trying to trace the registered keepers of allegedly insured vehicles was that at least 50% of enquiries to the vehicle register disclosed out of date or wrong information, although these enquiries may not have been representative of the vehicle register entries as a whole.

Consultations with stakeholders revealed serious problems in this area that appeared to be systemic. They were not, in other words, due to any fundamental inefficiency within DVLA or other agencies, but were due to the existence of an insecure system, which inevitably led to inaccuracies in the DVLA and other databases. According to DVLA, the main area of weakness was in the inaccuracy of the name and address of the keeper, particularly for older vehicles. Obviously it is vital for the accuracy of the vehicle register that the keeper must be identifiable. Moreover, it is these vehicles that are most likely to be untaxed, uninsured, without up to date MOT, stolen, involved in crime or abandoned. However, there was no real incentive for keepers to notify authorities about keepership changes. Although sellers and buyers were required to inform DVLA when the vehicle changed hands, many buyers failed to do so either to avoid paying excise duty (and other requirements including insurance, MOT and road traffic prosecutions) or to postpone payment for as long as they can. The incentive to avoid registration has become even greater with the advent of road congestion charging in central London. Vehicles entering central London are required to pay a "congestion charge," and their number plates are read by roadside cameras to check that such a fee has been paid. If not, the number plate is checked with DVLA for the name and address of the registered keeper, to whom a penalty notice will be issued. Avoiding registration as the keeper of a vehicle, therefore, makes it difficult for the congestion charge to be enforced.

Sellers, on the other hand, had no incentive to inform DVLA they had sold their vehicle. Even if they neglected to inform DVLA (and were therefore still registered as the vehicle's keeper), they were not held responsible for the vehicle including subsequent fines, tax, etc.

One solution suggested by many stakeholders was, therefore, for the U.K. to move towards continuous liability or continuous registration, commonly used in other countries. This means that registered keepers (even where they are not the actual drivers) remain responsible for tax, and all car-related fines and charges until the car is registered with another keeper, the car is officially scrapped, or the car is reported stolen. In this situation, they are also liable for the car if it is subsequently abandoned. This led to our **first recommendation**, that registered keepers should remain liable for their vehicle until a change of keepership is notified to DVLA.

Using this system, it is in the sellers' interest to inform DVLA that they have transferred keepership to someone else. If this were implemented, the common excuse of "I sold it to a man in the pub whose name and address I don't have" should be much less easy to make.

There would need to be further clarification about exactly which things keepers remain liable for (for example there is some question about the need to include basic insurance), but unless there are specific reasons why something should not be included, the principle should be that keepers should be liable for as much as possible.

It is important to note that continuous registration is not the same as continuous taxation, which may be resented by owners of vintage cars and vehicles used only seasonally. The concept of continuous liability may have to be presented carefully to the public to avoid confusion between these two concepts.

This is not, however, sufficient. Even if DVLA were to be informed of a new keeper, there is no official check on whether that person exists or whether the given details are correct. This means that new keepers may easily provide false names and addresses in order to avoid future fines, taxes, etc. Our **second recommendation** was, therefore, that a buyer should show reliable proof of identity (identity card/photo-ID and proof of current address) when registering as a new keeper.

In terms of addresses, the accuracy of both the Swedish system and The Netherlands databases relies to a large extent on the fact that all citizens are required to register their address changes centrally. Incorrect or outdated address fields were one of the biggest sources of inaccuracies in the DVLA database. If it had a higher public participation rate, the U.K. Government initiative to allow the public to notify government departments of a change of address in a single transaction would help in this regard. Although this system is voluntary, and the potential for misuse clearly remains, it should reduce the number of database errors and allow the enforcement agencies better to target their investigations.

Clearly a national identity card would help meet this identity requirement. But in its absence, photo identification (such as a photocard driver's license), plus proof of address, will be the minimum necessary, as occurs now for the purchase of mobile phones.

Such a system is currently used in The Netherlands where, prior to a private sale (private buyer and seller), the seller gives the buyer the document proving that the seller owns the vehicle as well as a transfer document (with the seller retaining the technical document). The potential buyer takes these to a post office along with personal identification papers (such as a driver's license with a photo). Once the buyer has registered with the post office and can demonstrate this to the seller, the seller can hand over the car. A similar model could, we suggested, be developed in the U.K. Post office employees could update the register and check all relevant vehicle details on-line, such as confirming its registration to the seller, and checking vehicle tax, insurance, and MOT status.

Other types of transfers occur when vehicles are sold to or bought from (or between) motor dealers, or others within the motor trade. In The Netherlands, most motor dealers are authorised to register these changes on-line. This could be achieved in the U.K. if there were a single central database or virtual database (as suggested in recommendation 3 below). Similar to post office workers, authorised motor dealers could check identification and all relevant vehicle details on-line, in addition to updating the register. Those dealers who were not authorised would have to involve the post office, much like private keepership transfers.

There are many other types of transfer, including corporate buyers of vehicles (who own a large proportion of the vehicles on the road), and first registrations of new vehicles, imported vehicles, and kit and rebuilt cars. The validation of identification in these situations would need to be clarified further, but the essential principle of checking authenticity of identity should remain in all types of transfers.

Two other issues were identified in this area. First, if the system were to be significantly tightened up, as we proposed, then it is possible that people already living at the margins of financial viability would find it very difficult to avoid being pushed into running vehicles illegally or not at all. This is an important point, which would need further investigation, and to which we make additional reference in the section on implementation.

Secondly, it was suggested that it might be appropriate to levy a nominal charge at the point of re-registration. Tightening up vehicle transfer arrangements would involve the deployment of substantial information technology (IT) resources, as well as staff resources at post offices, motor dealers and other outlets where transfers could be registered. Consideration would therefore be needed as to whether it was desirable and publicly acceptable to levy a modest charge to cover these costs.

A further complicating factor to the system was the number of separate vehicle databases that exist in the U.K. These include:

- DVLA vehicle register (holds details of keeper and vehicle);
- DVLA driver's register (holds details of all licensed drivers);
- Motor Insurance Anti-Fraud and Theft Register (MIAFTR; holds details of all insurance write offs);
- Police National Computer (PNC; holds details about stolen vehicles);
- Motor Insurance Database (MID; contains policy information relating to motor insurance policies);
- MOT database (this is still in the development stage but will eventually have details of vehicles' MOT histories); and,
- the Trailer register that DVLA is establishing.

These databases are mostly updated and used independently of each other. The fact that information on each of the databases is not brought together systematically creates a substantial barrier to accurate vehicle data for three reasons:

1. It is not easy to see the up to date registration details, MOT information, and insurance arrangements at one time, making cross-checking difficult.
2. Multiple entry of items of information, apart from being inherently inefficient, makes mistakes and inaccuracies more likely. It is also inconvenient for customers. There is no link, for example, between the DVLA vehicle and the driver databases, and motorists have to advise each section separately of address changes. The Government initiative to allow the public to notify multiple agencies of changes of address in a single transaction would help, provided the initiative received high uptake from the public.

3. There were particular problems with slow updates of databases. For example, it could take up to two weeks for the PNC and the DVLA databases to be updated after re-licensing. Clearly, real-time, accessible records are necessary to enable police and others to carry out their functions.

These problems would be overcome by moving to a single database allowing the police and other authorized users easy access to all vehicle details. This would seem conceptually simple, save on multiple keying and reduce inaccuracies. An alternative would be to link the separate databases, creating a virtual database with each system interfacing with the other seamlessly and in real time. An advantage of a linked rather than a new single database may be that it is less expensive. Our **third recommendation**, therefore, was that the U.K. should move to a single or virtual database by 2004.

One benefit of a single or linked database is that all appropriate agencies could have immediate access to important information. The database could also be verified and updated each time the motorist/vehicle comes into contact with an official agency in connection with the vehicle. Only those involved in official transactions (involving licensing, re-licensing, or an MOT inspection) would be able to conduct data checks.

Automatic cross-referencing of records could enable details such as keeper, vehicle tax, insurance, MOT test, recalls and so on to be more easily checked and appropriate action taken when they do not tally. For example, when tax is being paid a check could automatically ensure that insurance and MOT details were correct. Although currently it is not possible to purchase a vehicle tax road tax disc at the post office without proof of MOT and insurance, an automatic checking system on-line would mean there would be no need to rely on paper documents. Our **fourth recommendation** was that key details (vehicle and keeper identification) be verified and updated at each relevant contact with vehicles/keepers.

All relevant agencies would need to be linked to the unified database to conduct their business and carry out updates. These include DVLA, police, motor traders, insurance agents, salvage firms, MOT garages, and currently, post offices. This would result in more efficient business, better quality of data, better customer service, and better crime prevention and detection. Our **fifth recommendation** was, then, that Police and other appropriate enforcement agencies should have secure, on-line access to relevant parts of the database, leaving an audit trail for data protection purposes.

Under these arrangements, the motor trade could assume responsibility for updating the records for all vehicles purchased and sold, and garages could work with the central database when carrying out MOT tests to input new information on-line. The statutory proof of the existence of motor insurance could cease to be a paper certificate but could become a database entry. This would stop the current practice of some drivers who cancel their motor insurance as soon as their tax disc is issued. In addition, this would make life easier for motorists because authorised dealers could key in details directly. This system is similar to that used in both Sweden and The Netherlands, and is compatible with DVLA's current policy to provide electronic links for its customers to facilitate the direct update of the vehicle record.

It is obviously extremely important that the central database is up-to-date and accurate so that enforcement bodies can easily identify vehicles without insurance. Clearly to achieve this, collaborating agencies would be required to update the database very quickly after transactions, ideally in real time (i.e., instantaneously), but at least within the time-scale necessary to enable the operation of efficient business. Our **sixth recommendation** was that the insurance industry and the motor trade should supply relevant information to the database electronically within one working week by 2004 and move to real time by 2007.

For most agencies immediate updates would not be difficult. Indeed, some are already working towards this. For example, the DVLA's current dealer re-licensing and registration project envisages keeper changes captured directly and immediately at source. However, for other agencies immediate updates would pose a considerable challenge, particularly in the short term. In particular, the supply of information to the Motor Insurance Database (MID) by insurers is currently very variable due to the differences in insurance distribution channels. Insurers who rely on a number of small brokers for business can take considerably longer than direct underwriters and larger intermediaries. Although the industry has set itself a target of updating the MID within ten days, the average time to supply information about new policies after their effective date is around 25 days. The most significant delays are caused by the time taken for brokers to notify insurers. For instance, the Lloyds market as a whole takes an average of 33 days; and some insurers take considerably longer. Currently, many insurance brokers are not on-line, making it difficult for them to update the database quickly. In addition, the contribution to a broker's profitability made by retaining collected premiums, pending notification of the insurer, acts as a disincentive to immediate notification.

Ways in which insurance brokers could be encouraged to speed up notification should be developed. For example, as is planned for The Netherlands, the registration agency could publish a league table of insurers who fail to meet their deadline. This is intended to pressure them into dealing with the agency on-line rather than through the multi-various methods currently in use—fax, post, email, telephone etc. This should encourage greater compliance with the current legal obligation for insurance companies to send information to the central database within 28 days. Not withstanding this, in 2001 10% to 15% were still late. In the U.K., positive steps may need to be taken to improve performance in this area, but a target of real time compliance by 2007 seems feasible.

A number of stakeholders talked about issues deriving from the fact that insurance attaches to the person rather than the vehicle. There are possible road safety benefits in this arrangement, but stakeholders also noted that this has made for certain difficulties in enforcing action against uninsured driving. Linkages between databases would be simpler if insurance were attached to specific vehicles, as is the practice in many other countries.

These details are complex and were outside the scope of our study, but our **seventh recommendation** was that the Government commission an independent review of these arrangements.

Problem 2: Insecurity of Vehicle Identification Systems

A number of stakeholders contacted during the project raised concerns about the security of current vehicle identification arrangements.[3] In particular, vehicle ringing,[4] vehicle cloning,[5] and the use of false number plates were salient problems for many stakeholders. In addition, a number of stakeholders were concerned with the use of forged or fraudulently obtained registration documents. Number plates are relatively easy to obtain, manufacture or transfer between vehicles. This provides an excellent opportunity for criminals to disguise or misrepresent vehicles for resale or to avoid detection by law enforcement authorities. There are a number of ways in which these opportunities would be reduced. These range from relatively simple modifications of the existing number plates to the implementation of more novel identification systems, which make use of available technology. This technology is diverse, including spray-on identification systems[6] and electronic vehicle identification (EVI), which offer considerable scope for recording information on the vehicle. There are many advantages to EVI, particularly with respect to improving the detection of speeding vehicles, and as a potential aid to immediate tracking of stolen vehicles, which should help to deter car-jacking. DfT is currently involved in research examining electronic vehicle identification.

In other countries the problem is addressed by much tighter controls over the manufacture and issue of number plates. For example, some have only one centralised number plate supplier. There is also work underway in other countries to improve the security of number plates through, for example, the incorporation of machine-readable chips. To replicate the more effective and developing systems in other countries would require tighter control over the manufacture and issue of new plates and the replacement of all current plates. This would be disruptive and expensive. It may also be unnecessary, as it appears that EVI technology will be sufficiently advanced to provide a viable replacement for the identification function of number plates in the relatively near future.

Even with the introduction of sophisticated EVI at some future date, there will still need to be an easily recognisable and memorable identifier on each vehicle providing visual identification to witnesses of crime or accidents. It will be important that this number or other forms of identification is verifiable by any electronic identification systems that may be introduced. When EVI is introduced, the format of number plates can be revisited.

Our **eighth recommendation**, therefore, was that efforts should be directed towards the introduction of electronic identification of vehicles rather than investment in tighter controls on the current number plates. However, a pan-European solution will be necessary. More specifically, we suggested that the U.K. Government should propose to the European Commission that work starts urgently to introduce a European system of electronic identification of vehicles. Although the Commission has now started to look at EVI, negotiating common standards and interoperability will take some time. However, it is believed that five years is perhaps a realistic time frame for its introduction.

Problem 3: Inadequate Enforcement

Enforcement of VED, MOT, and insurance requirements was inefficient and ineffectual. There was a very low probability of being caught when driving without tax, insurance, MOT, and without having the vehicle properly registered to DVLA. This was partly due to the low level of resources dedicated to enforcement of the system, especially by the police and DVLA. Our **ninth recommendation** was that an independent enforcement agency be established, funded out of increased tax and insurance receipts.

For example, where the tax disc is out of date, police may issue a ticket requiring the driver to produce insurance, MOT, and driving license documentation at a police station. Although non-production of documents or production of out-of-date papers results in prosecution and the recovery of arrears of excise duty by DVLA, the system consumes a great deal of police time and does not always have a high police priority.

There was also a particular problem with enforcing insurance payment in the U.K. For example, it was possible for an uninsured driver to drive away from court after sentencing still having no insurance. There is also a perverse incentive with no penalty for a period of uninsurance (this is in contrast to the recovery of any unpaid road tax). To be proactive in following up uninsured motorists will require a designated agent because the police do not have the resources. It could not be left to individual insurance companies as they will not know whether a vehicle is insured elsewhere.

An independent, dedicated, self-financing enforcement agency to address licensing-related transgressions and breaches of the registration system would overcome these difficulties. The exact scope of this enforcement agency would need to be considered and there are a number of options:

- It only deals with insurance issues.
- It includes evasion of vehicle excise duty and MOT in addition to insurance.
- In addition to 1 and 2 above, it also investigates anomalies or discrepancies in the central database to improve detection of more serious vehicle crime.

The agency would need on-line access to the unified database in order to identify uninsured and untaxed motorists. It might also have powers to extract fixed penalties in order to minimise court time. Such penalties would be in addition to arrears payments for vehicle excise duty, MOT, insurance, etc.

The Netherlands and Sweden have comparable systems. The Dutch police are not responsible for chasing up unpaid vehicle excise duty. This is initiated automatically through the central database computer and it is followed-up by a private company not directly linked to the police. This company enforces tax, fines, insurance, etc. It acts as a debt collector and is able to go to the person's home to collect outstanding fines. Ultimately, if fines are not paid, the company is able to have the car towed and sold, with the money from the sale going towards paying the fine for the vehicle.

A similar system is used in Sweden. For example, if records show that a car is on the road but uninsured, a body under the auspices of the insurance industry is required to follow-up and secure insurance cover, in addition to collecting a punitive premium for the period since previous insurance expired. If the owner does not take out insurance, it becomes a matter for the police to impound the license plates/car.

One advantage of using an approach such as this in the U.K. is that police resources could be directed away from relatively less serious problems such as tax, MOT, and insurance evasion. However, strong links would need to exist between any external enforcement agency and the police as it is increasingly clear that tax, MOT, and insurance evaders are often those who commit more serious crimes. Furthermore, there appears to be limited public support for non-police agencies acting in an enforcement capacity.

Problem 4: Lack of Strategic Overview

If the recommendations made were to be accepted in whole or in part, then resources would need to be devoted to ensuring their implementation. Furthermore, it became clear during the course of the work that there is a great deal going on across various parts of government and the private sector in relation to vehicle registration and licensing. There are also emerging technologies, which could be used to great advantage in improving and securing the system. In 2000, this work was not well coordinated and there was no clear long-term goal, nor strategic overview of activities. These two tasks—implementation oversight and future strategic planning—needed to be addressed. Our **tenth recommendation** was that this could best be done through the establishment of two groups, an Implementation Group and a Futures Group within DfT.

The task of the Implementation Group would be to tackle existing problems, and its work to date is set out in the next section. The task of the Futures Group would be to identify how to minimise *future* social problems, including crime and social exclusion, particularly those that may arise from new technology. It is known, for example, that crime often follows when new technologies, products, environments, and systems are designed without such social consequences being thought through. New technologies can also facilitate enforcement activity. There will also be an ongoing need to ensure that new policies do not have perverse and unintended social consequences. For example, a fool-proof system of vehicle regulation may push some families into law-breaking, or make it impossible for some urban-dwelling individuals to support elderly relatives. At present, there is little central government effort directed towards anticipating these future threats and opportunities, and proactively incorporating an understanding of these into planning for the vehicle registration and licensing system.

The work of a Futures Group should focus on the following four areas:

- unambiguously linking vehicles and keepers;
- technology and vehicle identification;
- ensuring enforcement; and,
- encouraging inclusion.

It should be chaired at a high level within the department. By putting crime on the agenda at the highest level, it will ensure that it is considered at all levels. In addition, this group may help shape future EC directives and to defend against directives that might move in unwelcome directions.

Problem 5: Crime Prevention as a Priority

A number of stakeholders, in particular the police, suggested that DVLA might benefit from having crime prevention as a higher priority in the organisation. Police stakeholders identified a number of loopholes in DVLA procedures, which criminals could exploit. In particular, the vehicle registration document could be forged. As this is the only check that most purchasers will make (to ensure the details match those of the car), a forged registration document will allow a stolen vehicle to be returned to the marketplace. However, many criminals do not need to do this, as they are easily able to obtain legitimate documentation fraudulently. There are a number of ways in which this can be done:

- registering someone else's vehicle at your address;
- claiming vehicle is a grey EU import, i.e., has been imported personally rather than by a dealership;
- use of salvage to ring vehicles;
- use of stolen parts;
- claiming there has been some change to the specification of the vehicle; or,
- fraudulent use of the application form.

DVLA is working to close these loopholes but, despite improvements, criminals will always try to find ways to abuse the system, and so an increased understanding by DVLA staff (at all levels) may help them to detect suspicious data.

One issue identified by stakeholders as perhaps contributing to the problem is the focus by DVLA on customer service. Whilst customer service is very important, there can be situations when the need for providing a high level of customer service can actually be at odds with the need to prevent crime. If crime prevention were of higher priority on the DVLA agenda, this would be reflected in staff training and in the performance regime under which more junior staff worked. This would lead to greater awareness of "suspicious" data (such as change of name three times in one week) and the establishment of automatic electronic markers to flag vehicles, which the computer instantly recognises as "suspicious." There is, however, an important caveat here. There is little point in DVLA staff identifying suspicious computer entries if there is no adequate enforcement agency to follow up. This reinforces recommendation 9, which calls for a self-funding enforcement agency. Our **final recommendation** was, therefore, that crime prevention should take higher priority in DVLA.

All eleven recommendations are listed in Table 2.

IMPLEMENTATION

The JDI report did not land on a "greenfield site." Some of the recommendations in our report were already in the process of being developed in

Table 2 Eleven Recommendations for Modernising the U.K. Vehicle Licensing and Registration System

Problem: Database inaccuracy

1. Registered keepers remain liable for their vehicle until a change of keepership is notified to DVLA.
2. Buyers to show reliable proof of identity (identity card/photo-ID and proof of current address) when registering as a new keeper.
3. Move to a single or virtual database by 2004.
4. Key details (vehicle and keeper identification) verified and updated at each relevant contact with vehicles/keepers.
5. Police and other appropriate enforcement agencies to have secure, on-line access to relevant parts of the database, leaving an audit trail for data protection purposes.
6. Insurance industry and motor trade to supply relevant information to the database electronically within one working week by 2004 and move to real time by 2007.
7. Her Majesty's Government (HMG) to commission an independent review of insurance arrangements with respect to whether the person or the vehicle should be insured.

Problem: Insecurity of Vehicle Identification Systems

8. HMG should plan to introduce electronic vehicle identification before the end of 2007.

Problem: Inadequate Enforcement

9. An enforcement capability should be established funded out of increased tax and insurance receipts.

Problem: Lack of Strategic Overview

10. A Vehicle Licensing Implementation Group and a Vehicle Licensing Futures Group should be established.

Problem: Crime Prevention as a Priority

11. Crime prevention to take a higher priority within DVLA.

some way. For example, much had already been done to unify the various databases, as follows:

- Manufacturers and motor dealers now set up vehicle records under the Automated First Registration and Licensing System.

- DVLA was developing a system to enable motor dealers to notify DVLA of movements of vehicles within the trade. However, under current arrangements the system will be a voluntary one and motor dealers were showing little enthusiasm in participating. A regulatory change would be required to make the requirement mandatory.
- A system was also under development to enable motor dealers to re-license in-stock vehicles and notify DVLA of change of keeper details.
- DVLA had developed an electronic system, which allows salvage dealers to update records of scrapped vehicles. This system will be extended to enable compliance with the End of Life Vehicle Directive and it is anticipated that some 2,500 organisations will be linked up.
- DVLA was developing a link with the MIAFTR. This will update the vehicle register with details of seriously damaged vehicles.

Implementation is complicated, therefore, by the fact that our recommendations would need to be taken forward in the context of ongoing activity. This would inevitably require judgements and decisions having to be made about whether current activity is moving in the right direction or not and how to handle any anomalies. It may also require a greater sense of urgency and a clearer sense of direction. As an example, the government and the private sector are investing heavily in automatic number-plate recognition systems for traffic management, enforcement and, ultimately road charging. These systems rely at present on the integrity of the vehicle's number plate for identification. As we have already noted, the number plate is probably one of the least secure elements on the vehicle. It is more difficult to remove the external mirrors.

Additionally, more coherent and integrated system changes would be required. The recommendations were not always isolated from each other. There is no point, for example, in DVLA being able to identify potential infringements if there is no subsequent enforcement. With the current pressures on the police service to deliver major reductions in volume crime, the chances of extra resources being given to what are perceived as relatively minor traffic violations have to be seen as slight. The recommendation for a separate enforcement agency, staffed by individuals with specific training in the identification of suspect vehicles (which many serving police officers do not have), is, therefore, a crucial part of an integrated package.

Recognising these challenges, the JDI report recommended that an Implementation Group be established. This group should provide a high level forum for bringing together the different agencies, elements of the system and strands of work under a single framework, to ensure effective use of resources and focus on shared outcomes and a vision of the future. It should provide clear and unambiguous commitment from the top, without which nothing will happen. The JDI report identified the following specific tasks for an Implementation Group:

- Set targets in respect to implementation goals.
- Develop a marketing strategy to sell the ideas to the public.
- Co-ordinate implementation efforts across Government.
- Provide feedback to various interested parties.
- Check on the consequences of the actions.
- "Sweep up" outstanding work, for instance by looking further into social exclusion issues.

The Implementation Group

The eleven JDI recommendations were quickly accepted in principle by the Government, and an Implementation Group was indeed set up, chaired by a senior DfT civil servant. Called the Modernising Vehicle Registration Implementation Board (MVRIB), its members included the DVLA, police, vehicle insurers, vehicle organisations such as the RAC and AA, and other relevant Government Departments. Its terms of reference were to "Drive forward a programme to modernise vehicle registration in the U.K. to reduce vehicle crime and enhance compliance with road traffic law." A number of sub-groups were established to take forward more detailed implementation work on, for example, continuous registration, electronic vehicle identification, and motor insurance. The role of the main group was to ensure co-ordination and coherency across the programme as a whole.

The detailed workings and outputs of that group are beyond the scope of this paper. However, as outsiders to the process we able to observe the way a number of key issues affected implementation, that are worth highlighting.

The Impact of "Outside Events"

One of the important issues to manage in any large-scale implementation programme is the influence of outside "events." These are big changes in the implementation "environment," which can knock a programme completely off its intended course if unplanned for (see, for example, Homel et al., 2004). Such events can be unexpected, which makes them difficult to manage, and highlights therefore the need to be alert to the implementation context to try and anticipate such events better. This case provides an example where such alertness enabled an opportunity to be recognised and taken, with the result that implementation speed was greatly enhanced.

The case for continuous registration (CR) was a powerful one, since the success of the whole programme depended on having a reliable and accurate database of vehicles and keepers. Its introduction required, however, legislation and the JDI had been asked to avoid solutions that depended on primary legislation since this can be difficult to achieve in a very busy Government parliamentary programme where Departments compete with each other for parliamentary time. The DfT realised, however, that there was an opportunity quickly to pave the way for continuous registration, by "piggy-backing" on the legislative time provided for the Treasury. Every year the Treasury has a Finance Bill, to introduce any legislation arising from its budget. The DfT was able to include in that programme the necessary legislation for continuous registration since this was clearly a revenue issue, thereby avoiding the need for a separate departmental Bill. Having got the new legislation onto the statute book, secondary legislation was then required to establish the regulations about how CR was to work. This tight timescale gave the newly formed Implementation Group a focus and sense of urgency—the details had to be worked out and agreed within one year.

Dealing with Conflicting Priorities

The U.K. system is very user-friendly in that minimal bureaucratic effort is required to re-licence, buy and sell vehicles and inform DVLA of keepership change. Compare this with, for example, the U.K. system for buying, selling and registering change of ownership of real estate. It was important to the Government that this user-friendliness was retained as much as possible, and that the modernisation programme should not result in too much "red tape" and inconvenience for the general public. The recommendation proposing ID checks on new buyers, to ensure reliability and accuracy of new keeper details, presented a problem in this respect.

It was decided, therefore, to take a slower approach to this particular recommendation in the hope that the need for more detailed and costly identity checking could be avoided. Sellers are still required to notify DVLA of the name and address of the new keeper, and DVLA to maintain a watch on the names and addresses of new keepers that look fictitious or suspect in some way. In addition, the introduction of continuous registration was accompanied by an £8 million publicity campaign to explain how the new liability worked, and letters issued to keepers to remind and encourage them to re-licence. The impact of this approach would be monitored, and the need to introduce tougher ID checks reviewed at a later date.

It remains to be seen whether this milder approach will be sufficient. Previous research is not very supportive. The way in which a customer-service ethos can compromise effective enforcement and compliance, for example, has been documented elsewhere in the context of waste control (Webb & Marshall, 2004). The experience of introducing steering column locks in vehicles is that a similarly gradual approach to implementation taken in the U.K. had a much weaker impact on theft of vehicles than the "big bang" approach taken in Germany where all vehicles were required to be fitted with such devices within a short space of time (Webb, 1994). The evidence of the effectiveness of this "rule-setting" approach to date is unclear. While licensing activity has increased, there has also been a large increase in the number of keepers declaring to DVLA that their vehicles are not being kept or used on the road, such declaration releasing them from their liability to pay tax. Some of these declarations will be fraudulent, with keepers looking for ways to avoid continuous registration liability. Some, however, will be the result of the publicity campaign making keepers aware of the need to make such declarations if their vehicles are off the road. Clearly, the situation needs to be closely monitored.

Using Evidence

While there is much talk about the need for, and value of, evidence-led policy and practice, the reality is that there are few good large-scale examples of this working in practice. Some of the reasons for this have been explored in depth elsewhere (e.g., Homel et al., 2004). These include the timely availability of evidence, in a form that can be translated easily into policy and practice and in a context where evidence is valued and change is outcome-focused.

All these conditions were met here, which may explain why the project and the subsequent report were taken up so readily as the basis for the programme of change. The original JDI report, for example, was completed within 20 weeks of commissioning. While the problems were already recognised, the value for the Government was the way that it put together the evidence in a holistic way, and presented it and some imaginative proposals within a realistic policy context.

The senior civil servant chairing the Implementation Group had a background in the Office for National Statistics. He head-hunted another civil servant to manage the day-to-day business of the project, who also had a good deal of experience from the Cabinet Office in using evidence to deliver outcomes. These two key officials had, therefore, a healthy respect for research evidence and important experience in working with it on policy issues.

An important initiative to keep the Implementation Group focused on evidence and outcomes was to commission a series of follow-up research papers from the JDI. These were to be focused on the six priority problems of theft of motor vehicles, VED evasion, abandoned vehicles, driving without insurance, driving with no MOT, and driving whilst disqualified. The aim of each paper was quickly to pull together all the current relevant evidence and data to provide a rapid assessment of the scale of the problem currently, the likely impact if all recommendations were implemented, and propose suitable indicators for measuring impact.

In the event, papers were produced for the first three problems. Driving without insurance was dealt with by a separate and large-scale review of the vehicle insurance system, recommended by the JDI, commissioned from Nottingham University (Greenaway, 2004). There was insufficient data available to produce meaningful papers for driving without MOT or whilst disqualified, so these were dropped. An additional paper was commissioned, however, to make some assessment of the size of the so-called "vehicle underclass," those vehicles which are persistent and long-term evaders of the registration and licensing system, and a key target group.

These papers (Jill Dando Institute 2002, 2003b, 2003c, 2004) had a significant effect. Feedback from members of the group was that the continual presentation of evidence, packaged in a policy-relevant way, and showing what could be achieved if all the measures were implemented, helped keep the group focused and fast moving. As one member of the group put it, it is easy to lose sight of what you are trying to achieve when bogged down in the detailed implementation issues. More specifically, these papers helped decisions about target setting, led to the development of data-collection exercises for monitoring outcomes, and identified likely future problems in need of some urgent attention now.

As an example of future problems, the vehicle theft paper identified a reversal, albeit small but persistent, in the downward trend for thefts of relatively new vehicles, and showed how this could feed through to reverse the downward trend for vehicle theft as whole. Research is now underway in the Home Office to explore the problem of car keys stolen during house burglaries, as it is suspected this new *modus operandi* may in part be responsible for this upward trend. Another example is the abandoned vehicles paper which drew attention to the likely impact of the EU End of Life Vehicles Directive on abandoned vehicles in the run up to 2007, when producers take responsibility for disposal. The increased costs of disposal in the interim period would, it was estimated, at least double and possibly treble the number of cars dumped on the roadside. The Office of the Deputy Prime Minister has since drafted an abandoned vehicles strategy.

Financing the Programme

A considerable amount of effort was required from the Implementation Group to reach agreement with the Treasury about the financing of the implementation plan. Our initial assessment was that the arrangements proposed should reduce tax evasion by at least 50% (guaranteeing additional revenue of some £75 million per annum). In the first year these additional VED receipts would cover the costs of introducing the new system and thereafter there would be a net benefit to the exchequer of the order of £50 million per annum. We also expected a substantial increase in insurance premium income.

Treasury were unable to provide extra money to finance the cost of all the staffing, IT and advertising required to launch the scheme. Instead, they agreed a "netting-off" arrangement which allowed the DfT to retain fine income and set this off against start-up and operational costs. As luck would have it, and as we noted earlier, the scheme was more successful than forecast, with more motorists than expected re-licencing and re-licencing on time, bringing in some £75 million extra in vehicle excise duty—good news for the Treasury. At the same time, however, fewer motorists than expected incurred fixed penalty fines and, of those who did, fewer than expected paid their fines promptly, creating a significant cash flow difficulty for the DfT. The DfT are currently in discussion with the Treasury and others on the options to improve cash flow in order to maintain the considerable improvement in tax revenues that the scheme has brought about.

The Future

Our vision was that a secure and reliable database should form the linchpin of the new system. It is imperative that the vehicle is uniquely identifiable and linked unambiguously to the keeper, through the database, with real-time access for enforcement agencies. This is an achievable objective for new vehicles within five years and for all within ten.

Address correspondence to: Barry Webb, Jill Dando Institute of Crime Science, University College London, Brook House, 2–16 Torrington Place, London WCIE 7HN, UK (e-mail: b.webb@ucl.ac.uk).

NOTES

[1]The work arising from this consultancy is entirely the responsibility of the Jill Dando Institute. The views expressed are those of the authors and do not necessarily reflect the views of the DfT or any other government department.

[2]The steering committee comprised representatives from DTLR, DVLA, the insurance industry, The Jill Dando Institute of Crime Science, and Nick Ross, journalist and broadcaster.

[3]The three vehicle identification systems that currently exist are registration marks on number plates, VIN and the V5 registration documents which contain an official record of registration marks, vehicle identification numbers and vehicle and keeper details.

[4]Vehicle ringing refers to the practice of replacing the identity of a stolen vehicle with one previously written-off (usually done by switching number plates and in some case the VIN).

[5]Vehicle cloning refers to the practice of disguising the identity of a stolen vehicle with another of similar make and model (usually done by obtaining a copy of the original vehicles' number plates and attaching them to the stolen vehicle).
[6]http://www.mdatatech.com/whatsnew.htm

REFERENCES

Burrows, J. (1991), *Making crime prevention pay: initiatives from business.* Home Office Crime Prevention Unit paper 27. London: Home Office.

Greenaway, D. (2004), *Uninsured driving in the United Kingdom.* A report for the Secretary of State for Transport. London: Department for Transport.

Homel, P., Nutley, S., Webb, B., & Tilley, N. (2004). *Investing to Deliver: Reviewing the implementation of the UK Crime Reduction Programme.* Home Office Research Study 281. London: Home Office.

Jill Dando Institute. (2002). *Measuring the impact of MVRIB initiatives on vehicle theft.* London: JDI (available on www.jdi.ucl.ac.uk).

Jill Dando Institute. (2003a). *Government regulations and their unintended consequences for crime: a project to develop risk indicators.* Final report to the EU Crime Proofing Steering group. London: JDI (available on www.jdi.ucl.ac.uk).

Jill Dando Institute. (2003b). *Measuring the impact of MVRIB initiatives on vehicle excise duty evasion.* London: JDI (available on www.jdi.ucl.ac.uk).

Jill Dando Institute. (2003c). *Measuring the impact of MVRIB initiatives on abandoned vehicles.* London: JDI (available on www.jdi.ucl.ac.uk).

Jill Dando Institute. (2004). *Estimating the size of the vehicle underclass.* London: JDI (available on www.jdi.ucl.ac.uk).

Levi, M., & Handley, J. (1998). *The prevention of plastic and cheque fraud revisited.* Home Office Research Study 182. London: Home Office.

Martell Consultants. (2002). *An inventory of vehicle related data.* Perth, Scotland: Martell Consultants.

Newman, G. R., & Clarke, R. V. (2003). *Superhighway robbery: Preventing e-commerce crime.* Cullompton, Devon: Willan.

Smith, M., Jacobson, J., & Webb, B. (2004). Abandoned vehicles in England: Impact of the End of Life Directive and new initiatives on likely future trends. *Journal of Resources, Conservation and Recycling, 41*(3), 177–189.

Smith, M., & Webb, B. (2005). Vehicle Excise Duty evasion in the UK. In M. Smith & N. Tilley (Eds.), *Crime science: New approaches to crime detection and prevention.* Cullompton, Devon: Willan.

Webb, B., & Marshall, B. (2004). *A problem-oriented approach to fly-tipping. A report for the Environment Agency.* London: JDI (available on the JDI website: www.jdi.ucl.ac.uk).

Webb, B. (1994). Steering column locks and motor vehicle theft: Evaluations from three countries. In R. V. G. Clarke (Ed.), *Crime prevention studies* (Vol. 2, pp. 71–91). Monsey, NY: Willow Tree Press.

Security Coding of Electronic Products

by

Ronald V. Clarke
Rutgers, The State University of New Jersey

and

Graeme R. Newman
The University at Albany, New York

Abstract: *It is well established that certain products create opportunities for crime because they are useful "tools" for criminals or they lack security features that make them ready targets for theft. This paper takes the first step towards establishing "crime proofing" codes that assess the vulnerability to theft of one class of products: portable electronic devices. We begin by proposing a general framework for thinking about security codes, the main elements of which are: (1) corporate social responsibility, and (2) the economic arguments for regulating negative externalities produced by industry, of which product-related crime is one. This analysis leads to the conclusion that the most efficient form of regulation would be a voluntary code, administered by the electronics industry (specifically its trade associations), with some limited but crucial support from government. A draft security code is constructed based on two dimensions: (1) the intrinsic "hotness" of the product derived from previous research, and (2) the security features that have been built into the product or its marketing.*

Crime Prevention Studies, volume 18, pp. 231–265.

INTRODUCTION

This paper takes a first step to constructing a set of security codes for criminogenic products, i.e., manufactured items that are the targets or the tools of crime. The commissioning body for this paper, the Foresight Crime Prevention Panel (see Introduction), recognized the complexity of this task and suggested that the first step should be confined to electronic products since these play an important part in many crimes. Accordingly, this paper proposes a code that will establish performance standards for the electronics industry in theft-proofing its products. This code makes use of two checklists, the first of which is our adaptation of the CRAVED model of "hot products," i.e., those that are attractive to thieves. The second is a checklist we have constructed of product security features. We also outline the process for "security coding" of products and we recommend a series of steps that must be taken if the codes are to be brought into force.[1] These steps recognize that incentives must be put in place if businesses and manufacturers are to adopt security codes.

In developing the code, we considered it important that it should serve as a prototype for coding other kinds of products than those of the electronics industry. This required that we establish a general conceptual framework for designing such codes, and we attempt this in the first part of the paper. Of necessity, this discussion is undertaken at a general level, but we make reference throughout to electronic products.

PART 1: CONCEPTUAL BACKGROUND

In this section we consider, first, the nature of business responsibility for dealing with crime related to its products. We then move on to a brief discussion of the complexities of product crime risks and explain the considerations that guided us in narrowing our focus to theft of a specific group of portable consumer electronic products. Finally, we give an account, based on economic theory, of our treatment of product-related crime as an "externality," similar to pollution, and of our reasons for advocating a voluntary code administered largely by the electronics industry with some limited but important government support.

Corporate Social Responsibility

While once a controversial idea, it is now well established that certain products are criminogenic. Thus, they may lack security features that make them ready targets for theft or they might be useful "tools" that facilitate crime. In fact, manufacturers have done much to modify the design or packaging of products in order to prevent crime, and have done so for a variety of reasons, including consumer demand, legal liability, pressure from industry groups (e.g., insurance) and sometimes government intervention (Clarke & Newman, 2005, this volume, "Modifying Criminogenic Products . . . "). Businesses have generally moved fastest to modify products when they are the main victims, and have been much slower to act when the public is the main victim, unless sales are threatened of the products concerned. They are slowest of all to act when the modifications offer them little or no commercial advantage, when the crimes seem trivial (e.g., car break-ins or vandalism) and when they are scrambling to develop a new product, such as the mobile phone.[2]

These findings might lead one to conclude that business has little incentive to modify its criminogenic products (or practices) unless it is directly harmed by the resulting crime. But to assume that all businesses are unwilling to bear responsibility for the crime resulting from their products is a gross oversimplification. Indeed, one must recognize that the motives that drive businesses (and the individuals in charge of businesses) are complex and varied. In many cases, these motives include meeting an ideal of corporate social responsibility, which includes reducing the criminal harms resulting from deficient product security or design. As argued in a recent Institute for Public Policy Research (IPPR) report (see Hardie and Hobbs, this volume), businesses bear this responsibility because they are much better placed than anyone else to correct the security or design of their products, but as the report shows there are many other pressures on businesses to act in a socially responsible manner. They include pressure from special interest groups or the media, pressure from investors and from peers within the industry, the need to safeguard reputation and self-esteem, the desire to be a good citizen, a general sense that business should help solve public problems, fear of regulation, sometimes profit and, of course, compliance with the law.

The proposals made below for developing a security code for electronic products are predicated on the assumption that industry leaders are committed to an ideal of corporate social responsibility, even though this will be tempered by the usual requirements for profit and for freedom from burdensome regulation. Following the IPPR report, we have therefore tried to map out a middle ground that anticipates both government and business participation in constructing and implementing the code. The justification for this approach is found in an economic model (described in a later section of the paper) that treats crime as an externality in much the same way that it treats environmental pollution. This approach requires a rational approach to decision making that entertains the costs and benefits of designing security into a product, of introducing a code to establish a set of security performance standards, and of establishing incentives for compliance with the code.

We argue that to think of incentives as negative or positive, or as voluntary or involuntary, can be misleading. While the strategy we finally recommend may be loosely characterized as voluntary, in fact the incentives that are attached to voluntary adoption of codes may be positive or negative. Usually, they will turn out to be a mixture of both. If pressure comes from customers for a secure product design, is this a positive or negative incentive? Coming from the marketplace, the temptation is to call it positive. But if the same pressure comes from a government agency, the tendency is to call it negative. The logic of this position leads to the conclusion that only a business decision to design security into a product made by a corporation itself, for reasons other than external pressure, could be thought of positive. This is nonsensical, of course, because business decisions can never be made without due regard for all external conditions, first and foremost of which is the marketplace. The economic model of externalities provides a framework for assessing the advantages and disadvantages of various strategies of compliance, and especially a way of assessing the role of government in ensuring this compliance.

In fact, we believe that government has a crucial role in fostering industry cooperation not just with government, but also among its own trade associations in order to establish the security by design organization. Government could make an important input to the organization that collects information concerning product risk by ensuring an impartial and independent procedure for collecting the relevant information. This impartiality and independence is necessary given that there are competing interests among industry in regard to what information can be made public and the extent to which various marketing and product data can be revealed. The government will also need to assist in promoting public awareness of the importance of product security and civic responsibility for preventing product-related crime. Promoting public awareness of product security could help create customer demand for secure products as well as fostering a sense of individual responsibility for preventing crime that may arise as a result of owning a particular product. The governmental approach therefore should be one that conceives of a balance of civic responsibility: not only do businesses have a responsibility to prevent crime resulting from their products, but so also do individuals have a responsibility to make sure that their ownership of a product does not contribute to its criminality or vulnerability to crime.

The Complexities of Product Risk

Many products are the targets of crime, as in the theft of cash, the theft of videos in the course of burglary and the counterfeiting of luxury items. Products may also facilitate many kinds of crime. Guns are misused to commit robberies, prescription drugs are abused to satisfy drug addiction, and telephones are used to commit telecommunication fraud. In addition, services attached to many electronic products may also enhance the capability and effectiveness of traditional crimes, such as bank frauds committed by hacking into bank computers, or vandalism by disrupting a company's e-commerce web site (Newman & Clarke, 2003).

Crime risks vary widely according to the type of product, within product categories and across brands. For example, certain car models are at much greater risk of theft than others, and which models are stolen depends on the nature of the offence. Those taken for joyriding are quite different from those taken for resale, and both are different from models that are targeted for accessories or parts (Clarke & Harris, 1992).[3] The British Crime Survey shows that the objects most consistently stolen from private individuals include cash, vehicle parts and accessories, clothes and purses or wallets (Clarke, 1999). Annual surveys in the United States show that, while shoplifted items vary from store to store depending on stock, certain items are also consistently stolen more often. These include cigarettes and alcohol, designer apparel and training shoes, audio and video CDs and cassettes, trinkets and jewellery, and medicines and beauty products (Clarke, 1999). In general, electronic products are among the hottest products for burglary, and car radios are the most common product stolen from cars (Clarke, 1999). These variations in theft risk are captured by the acronym CRAVED. Thus, products that are at greatest risk of theft are **C**oncealable, **R**emovable, **A**vailable, **V**aluable, **E**njoyable and **D**isposable (see further below).

A further complication is that a product's attractiveness to thieves depends not just on its intrinsic properties, but also on some external factors. One of the most important of these is the product's novelty. Many products are at little risk of theft when first introduced to the market because they are unfamiliar and little in demand. Later, they become much more prone to theft as more people begin to want them. Risk of theft may also vary with the stage of the product in the supply chain, with the highest risks occurring at points of transfer such as shipping and receiving, and in the stock rooms of retail stores. These well known points of vulnerability provide opportunities for employee theft.[4] At this juncture, the risks of theft move, first, to shoplifting from the sales floor and, finally, to theft from the legitimate purchaser. Once the customer takes ownership, the distribution of risk of theft changes according to where the product is (in the car, on the person, in the home) and the risk also changes as the product ages. In most cases, newer products are at greater risk, but some car models are more often stolen when they are older. What is essentially involved here is the distribution of risk according to ownership of the product. The security response to this problem would therefore seem to lie in product design and/or packaging that ideally included a means of easy authentication of ownership as the product passes from one stage to the next in its life cycle, such as chipping and radio-frequency identification tags (RFIDs).[5]

To find a way through these complexities, we decided to simplify our task by developing a prototypical code that could, at some later time be extended and adapted to a much wider range of products and crimes. Our first task, therefore, was to reduce the range of products and possible crimes. The Foresight Crime Prevention Panel, for whom this study was originally conceived, had begun this task for us by limiting the study to electronic products, but the range of such products and possible crimes is still quite extensive. And for reasons discussed below, we decided to confine our prototypical code to deal only with theft of portable electronic products, specifically those serving communication, entertainment and personal productivity needs.

THE SCOPE OF THE CODE

The tender documents for this project called for the construction of a security code to establish voluntary standards at manufacture, which would identify the "criminogenic capacity" of any "electronic products and the services directly related to these." In keeping with the "futures" orientation of the Foresight Panel, we reviewed a bewildering variety of new and anticipated electronic products in the information technology field, such as embedded linux devices for controlling home appliances, Internet radios and audio systems, cordless-webpad/phones, mobile multimedia communicators of various kinds, mobile video phones, TIVO personal video recorders, mpeg car radio players, wristwatch web pads, mobile and cordless network servers, and personal satellite assistants. Which of these will become products of mass acquisition is extremely difficult for marketing experts to predict, let alone for the authors of this paper. Furthermore, any set of codes should take into account the rapid change that is typical of the information technology environment. This would be especially important if "hot services" directly related to electronic products were to be included as products in any such set of codes.

The challenge facing us, therefore, was to construct a code that was not product-specific, but which nevertheless was sufficiently concrete to be implemented with little further development. The code also had to be capable of being extended to cover other products (and crimes), otherwise it would be necessary to start from scratch every time a security code was needed for a new category of products. Finally, the code had to be independent of any anticipated changes brought by developments in technology, or else it would be under constant revision to take account of the constant changes that accompany information technology. Our solution to this daunting task was to construct a general framework for thinking about security codes for products and within this framework to begin at a very simple level by constructing a code for a restricted range of products and crimes. As noted, Foresight began this simplifying process for us by limiting the code to electronic products and the services directly related to them. We have found it necessary, however, to limit the scope of the code even further as we discuss below.

1. Only Consumer Electronic Products

A simple, though important distinction, between consumer products and business products helps to reduce the range of products considered in developing our framework for a code. We have considered a number of ways of classifying products in order to assess their criminogenic properties (Clarke & Newman, this volume, "Modifying Criminogenic Products . . ."), but have found the distinction between business and consumer products the most sustainable, though also not as clear cut as one would like. Consumer products are those products owned and maintained by individuals for their own use. Business products are ones that businesses own and use in order to market, sell or maintain various consumer products and services. These businesses include financial institutions, transportation and telecommunication companies, public institutions and government bodies (when acting as service providers). Useful as it is, this distinction is blurred by the fact that private individuals regard some products that are owned by business, such as credit cards, as their personal possessions when in fact they are essential tools of business. The more typical business products are automatic teller machines (ATMs), cigarette and other vending machines, cable TV boxes, and satellite dishes. It will be noted that vending machines typically dispense simple consumer products for personal use. ATMs, cable boxes and satellite dishes dispense services of various kinds. One product that falls somewhere in between business and consumer products is the telephone, particularly the mobile phone. At one time, most home telephones were rented as part of the service from the telephone company, so could be termed essentially business products. Today, these devices may be bought and owned by the consumer. However, in the case of mobile phones, many marketing schemes give away the device as a part of the telecommunications service. One might therefore argue that credit cards and mobile phones are similar consumer products. However, in the case of the credit card, it is not possible at the present time to purchase a credit card in order to own it.[6] Mobile phones, especially those with additional features, do have some intrinsic value as devices, but their major value lies in the *access* to the services that these devices provide.

Thus, we include mobile phones in our limited list of products, but exclude credit cards, as we also exclude cable TV boxes and satellite dishes. We do not deny that some of these products are targeted by thieves for their services, especially cable TV boxes, but for reasons of simplicity, we consider it necessary to confine the development of the initial code focused only on consumer products, narrowly defined. This approach is also consistent with the Foresight Panel's focus on the consumer and the consumer's product as the ultimate key to establishing a more widespread public understanding of the costs of crime, and ultimately to creating a demand among consumers for crime-free products. If a consumer's product is stolen, the consumer bears the burden of loss. If a business product is stolen, by and large, the business bears the burden of the loss, though the loss may eventually be passed on to consumers as a whole.

2. Not Large Appliances

Appliances that are not easily removable, such as ovens, washing machines and dryers, refrigerators and freezers, and large specialty audio systems are excluded because we know from previous research that these items are rarely the targets of theft, simply because they are too big to remove easily.

3. Products as Targets not Tools

Any object or service could conceivably be used to commit crime of some sort. As an extreme example, an individual could use a VCR as a weapon with which to hit someone. There is no conceivable way that products could be designed in order to prevent all misuse.[7] As noted, some consumer products are stolen not for themselves but as tools to obtain access to the services attached to them. We argue here that, although these products are stolen to serve as a tool to obtain access to valued services, it is the access that is key. In the example of the VCR used as a weapon, any hard object that could be lifted to strike a blow could serve the same function. But in the case of gaining access to the range of services attached to mobile phones, the mobile phone or its equivalent must be stolen or otherwise acquired. Thus, it is the service attached to the telephone that makes the telephone hot. This leads us to our next step in simplifying the range of products to include in our code.

4. Not Services as a Primary Target

All indications are that in the future services will become a dominant force among consumer products. *Turning the Corner* (Wall & Davis, 2001) identified "hot services,"[8] and the Foresight Panel repeated this idea as part of the tender document for a crime-proofing code. In our review of the variety of consumer services and their relationship to consumer products, we uncovered a bewildering array of services that are constantly changing, and which have highly complex relationships with particular products. The variety and complexity of the criminogenic properties of information and information systems that are the prime components of services—indeed are often products in themselves (e.g., banking and investment products such as currency transfers or online stock purchases)—we have examined in detail elsewhere (Newman & Clarke, 2003). Suffice it to say that the expanding information society clearly means that the element of service will undoubtedly assume an ever-increasing part in regard to most future products. Many products, it is expected, will become "electronicized" by new service features. The best example of this change is the current push towards devices that control many appliances in the home, from security systems, heating and cooling, to kitchen appliances[9] and entertainment centers.[10]

However, for the purposes of the present task, we find it necessary to exclude services as a product in themselves because of the special attributes that they carry, and the impossibility of defining their characteristics independent of the complex information systems that both contain and transmit them. We have already noted that services *are* products in their own right. We are also very mindful of the fact that the information environment entails special crime-inducing features. Indeed, we have argued in our book *Super Highway Robbery* that information may easily be conceived of as a hot product, perhaps *the* hot product of the 21st century (Newman & Clarke, 2003). Its analysis and assessment in terms of codes, however, would take us on a far-reaching examination of the crime-inducing features of information systems, which make it possible for services to be delivered to customers. The design of information systems from the point of view of security and crime prevention is an extensive topic of its own, and one that we consider impossible to deal with given the task at hand (Anderson, 2001). It is a topic that we consider pressing, but outside the range of the present task, and the need on our part to keep to as simple a model for a framework as possible.

5. *Only Theft, not Misuse or Abuse*

The Foresight tender document called for a code to prevent not only theft, but also product misuse and product damage. As noted in point 3 above, the question of product misuse is beyond the scope of the present exercise. In the case of mobile phones stolen to access services, we do not class this as misuse. It is the proper use of the phone for what it is manufactured to do, but by the wrong person. Whether or not the thief will use the phone to gain illegal use of phone service is again not a question that we need to answer. By focusing only on the object that is stolen, we need not get into the extensive ramifications of the services to which they provide access—essentially those of telecommunications and the enormous variety of Internet services. We also consider the question of product damage, though important, to be an unnecessary complication at this stage. It is evident that some criminogenic factors of products include their vulnerability to damage or vandalism. This may often have to do with the packaging of the product, which may be designed in order to prevent theft, but also prevents the direct access to the product by a consumer when shopping. We refer here to the literature on product tampering, discussed in our paper prepared for the Home Office (Clarke & Newman, this volume, "Modifying Criminogenic Products . . . ").

6. Not Counterfeiting or Illegal Copying

A difficult category of crime, related both to packaging and to damage of products, is that of product counterfeiting. Various methods such as holograms and microdot printing are used both on products themselves and on their packaging in order to prevent counterfeiting. To include counterfeiting in this exercise would open up a host of additional problems. For example, it would be necessary to consider dual cassette recorders and CD/DVD burners as special items enhancing counterfeiting because these products make it easy to make illegal copies of videos and software. However, we must recognize that, except in unusual circumstances, consumers stand to gain financially from illegal copies whether they purchase them or make them on a device bought for the purpose. Thus, identifying the criminogenic properties of, say, a CD/DVD burner will make it difficult to convince consumers that buying or purchasing illegal copies contributes to the social cost of crime. And in regard to theft, it is unlikely that a counterfeiter will steal a CD/DVD burner for the purpose of making and marketing illegal counterfeit copies of well-known songs or software. Rather, this is the domain of "professionals" who invest in the necessary technology in order to produce high quality counterfeit products. So long as the article is of sufficient quality and the price is right, the consumer will not ask or care where the article came from. The consumer sees therefore no gain in supporting any public demand or movement to stop this type of crime. It may be argued that consumers do stand to lose if they buy, say a Rolex watch that is counterfeit, but this is true only if the consumer has been deceived into thinking he/she has bought an original. However, in the domain of luxury items, highly sophisticated criminal organizations invest considerable capital in order to exploit the brand names of luxury products. They could not profit from this without a strong consumer demand for a status product at a cheap price. The only recourse for the manufacturer (apart from not producing a brand name product at all) is to build in unique identifying features or, as is now common, to debase the brand name itself by producing cheaper versions of the product under the same brand name.

7. Only Physical Media

We include media as physical products that enhance the value of an electronic device. We do not include them as containers of music, software or video. We do this because of the complexities of accommodating the problems of music and video that are now downloadable from Internet sites, the difficulties in distinguishing between legal and illegal copying, and the extensive technical ramifications of the information environment such as software locks, encryption, hacking and other criminogenic features of the Internet (Newman & Clarke, 2003). Media such as disks, computer chips, ram (including "smartmedia" and "ram sticks" used for digital devices such as cameras) are included as physical products only, though some of these products may or may not be more hot than others if they provide access to the Internet or other services as do mobile phones. In other words, we see these media as being similar to batteries. They are needed to unlock the value of an electronic device. But unlike batteries, some types of media such as music CDs may contain information than is valuable in its own right. Access to that information may increase the hotness of that physical product, thus making it attractive to steal. We emphasize once again, however, that when we say "steal" we mean stealing the physical product, such as a CD, not stealing the music that is contained on that CD or elsewhere. If we considered the music contained on the CD as a primary target, we would have to also include the web sites on the Internet that make it possible to download the same music for free, in violation of copyright laws. And as we have said, we do not think it possible at this early stage to incorporate the complex criminogenic features of the Internet in our first framework for a code.

8. List of Products to Be Covered by the Code

Given the above constraints, the products to be covered by the code include all consumer electronic products serving goals of communication, entertainment and personal productivity, such as: personal computers, laptop or notebook computers, personal digital assistants (PDAs), personal recording devices, digital cameras/video cameras, GPS devices, phones/answering machines/fax machines, mobile phones, VCR/DVD players/personal TV players, TVs/remote controls, audio systems, radio/CD/DVD players, electronic musical instruments, game players, and personal media devices and their accompanying media. New or recent products would also include those listed earlier: embedded linux devices for controlling home appliances, Internet radios and audio systems, cordless-webpad/phones, mobile multimedia communicators of various kinds, mobile video phones, TIVO personal video recorders, mpeg car radio players, wristwatch web pads, mobile and cordless network servers, and personal satellite assistants.

THE CASE FOR A VOLUNTARY CODE

The role of codes—or standards of performance as we would prefer to call them—and the incentives needed to implement them are extremely complex in a modern economy. In order to understand this complex relationship it is useful to borrow from the thinking of economists concerning the costs of the actions of manufacturers to society, and the ways in which government may or may not be useful in contributing to incentives to affect the amount and distribution of such costs.[11] When a manufacturer takes an action that has an effect on another company or individuals for which the latter do not pay or are not paid, economists call this outcome an *externality*. There are a variety of such externalities and ways to accommodate them. When crime results partly from a manufacturer's design and/or marketing of a product, a negative externality has occurred, and it is regarded as a "social cost"—in contrast to the marginal cost of producing the product in the first place. It is also important to recognize that externalities may be produced by either producers or consumers. For example, if individuals make illegal copies of software and sell them on an Internet auction site, they create negative externalities for the software companies. This is extremely important when we consider that the long-range goal of the Foresight Panel is to create a public demand for products that are secured against crime.

Economists also think of externalities as resulting from the use or misuse of common resources, such as the air, or water. For example if there is a town commons available for all to use, it may become overgrazed or environmentally degraded by excessive use by individuals or business unless there is some mechanism in place to control or distribute access to it. It is possible to think of crime as a common resource that gives a market advantage to those who use it.[12] This is similar to pollution that gives companies a market advantage because it costs them less to produce their goods if they do not have to pay for pollution-reducing procedures in manufacturing their product.

Public versus Private Solutions to Externalities

Economists identify basically two approaches to the solution of externalities, which are public or private. It is apparent that the Foresight Panel anticipates a private solution, since the tender document requests a voluntary code. However, in its request for a plan to provide incentives for implementation of the code, the Panel remains silent on the role of government. It would appear that the government must play *some* role as far as incentives or implementation of a voluntary code are concerned, since the output of the Foresight Panel is directed not only to manufacturers, but also to other government departments, and anticipates that the Home Office will continue the work already started. We should begin at the outset by noting that the task that awaits us is extremely challenging. The obvious question in regard to implementation of codes is: why should a manufacturer implement design changes for a product if it will raise the price of the product and possibly place the manufacturer in a negative situation in regard to competitors who do not comply with a code?

The private solution to externalities may be of three kinds, or some combination of these:

- The first is known as internalization of the externality, in which the firms or individuals producing the externality get together and absorb the negative costs of the externality. In our case this would mean that the manufacturers should make an agreement among themselves to design security into their products so as to reduce the opportunity for crime. The way this would be done would be through a set of standards that all would agree on. We should note, however, that the internalization of the cost would mean that the added cost of producing crime-prone products would not be passed on to the consumer. This would have to be part of the agreement among manufacturers.
- The second kind of private solution is for the producers of negative externality to get together and work out a "deal" with those who are the victims of the externality. This would mean some form of compensation to them in lieu of the damage done by the manufacturers'

product. In our case, this could take the form of a "theft guarantee" in which the manufacturer (perhaps also in collaboration with the retailers) guaranteed to replace the item if it were stolen within a certain number of months after purchase.

- A third kind of private solution is civil litigation in which individuals or firms sue the producers of negative externality for damages done. In the case of crime-prone products, this has occurred in the USA in regard to handguns, and in the U.S. and U.K. in regard to product tampering (Clarke & Newman, op. cit.).

Proponents of the private solution to externalities claim that this is the most efficient way of dealing with the externality. It is a classic "free market" solution.[13] Unfortunately, history provides ample examples that cooperative agreements on their own have not fared well in dealing with negative externalities. There are four main reasons why some form of government intervention is needed.

1. There can be serious problems in arriving at a cooperative agreement among all manufacturers. In fact, the last manufacturer who holds out in agreeing to internalize the externality holds a great advantage, since he or she can produce the product without the cost of making it crime-prone, and so can take it to market at less cost than his or her competitors.

2. Information that is delivered to customers may not be accurate. In fact, it is not in the interests of manufacturers to collect and disseminate information concerning the crime-prone nature of their products. In other words, not only is there no economic incentive for them to make an agreement among themselves in the first place, but there is also an incentive for them not to inform the public or customers concerning the negative externality of their product. Or, given the power of advertising, it may even be possible for them to misinform customers in regard to their product.

3. The cost of getting the various organizations (manufacturers and retailers of electronic goods) together is not insignificant. It can be argued that government provides such organizational services as a public good, which is a form of internalizing externalities. This may, in fact, describe just the current role of the Foresight Panel.

4. The costs of litigation are extremely high, and the vagaries of accumulated common law in regard to civil suits related to questions of common resource are considerable, contributing to extreme inefficiency in settling negative externalities.

In light of the above, the advantages of government helping solve negative externalities are that: (1) it can absorb the transaction costs, since an additional organization to deal with each type of externality does not have to be established, and (2) it deals directly with the problem of the free rider who does not participate in the agreement to internalize the negative externality—in our case the problem of non-compliance with the established code. So that we can make a more informed comparison between pubic and private solutions to externalities, let us look more closely at the public remedies in the hands of government.

Mandated Codes

The government has four basic remedies available to it: fines, subsidies, regulations, and the legal definition of property rights. The last-named refers to the problem of common resources, such as oil, fishing rights and so on, so is not directly relevant to the problem of crime-prone products.[14] The first three, however, are remedies that are widely used by government in dealing with many externalities. For example, there is extensive regulation of health and safety in the workplace that is conducted by government departments. And in regard to products, there are complicated processes of regulation of many different kinds of products in regard to safety, e.g., prescription drugs, medical procedures and devices, food, automobiles, toys, tools, electrical products. The type and degree of regulation depends on the product. In addition there is extensive and complicated regulation concerning the negative externality of air and water pollution, most of which is under constant scrutiny by politicians and interest groups.

Before examining the three relevant remedies to regulating crime-prone products we should emphasize what may be obvious: these solutions assume that compliance is involuntary, or put more bluntly, standards or codes are imposed on the manufacturers, using one of the remedies above. However, economists always view these impositions in relative terms. The goal is never to completely wipe out, say, pollution. Economists regard this as an unrealistic goal. Rather, the costs and benefits of reducing pollution have to be considered just like any other economic activity. The same also obviously applies to crime. It is necessary, from an economic point of view, to compute the socially efficient level of crime. The argument stated in this way therefore points to the central question: do manufacturers take into account the social costs of the externalities they impose when producing a product that is criminogenic?

Fines (or taxes which amount to the same thing) should therefore be calculated based upon a computation of the social costs of the externality caused by the product.[15] In this way, the manufacturer is forced to consider the social costs of the product, or at least account for it in production and pricing. The problem with this remedy, however is that the extra cost of producing a crime-free product will most likely be passed on to the consumer, who thus pays twice for the externality—once in higher prices and once in income taxes paid to the government that regulates the industry. Furthermore, the computation of the amount of fine that equals the social cost is extremely difficult and to the extent that it is inaccurate, it will produce serious inefficiency, whether the computation is too high or too low. If too high, it may reduce output or production, thus raising prices even higher. If too low, business may mostly ignore it. And, of course, in order to make fines work efficiently, there is a high government cost of enforcement.

Subsidies do something similar, but in a positive way. The government may subsidize in various ways the technology or other method used to reduce the crime-prone feature of the product. Subsidies are usually preferred by business because they provide firms with the flexibility to develop their own technologies for designing crime out of products. They may manage to do this in a competitive way, even increasing profit margins. However, this solution can become complicated and highly inefficient if the government tries to dictate the processes or technologies the manufacturer should use. This is because the manufacturer can generally be assumed to have deeper knowledge of the technologies and techniques necessary to design its products, and should therefore be better able to judge what solution is best. By and large, the modification of products to design out crime supports this contention (Clarke & Newman, op. cit.). However, the upshot of this observation is that the only way that manufacturers can be judged as to whether they have introduced procedures to design out crime from their products is to collect detailed information concerning the incidence of crime—in our case theft—concerning the respective product. This is an extremely important point to which we will return shortly.

Meanwhile, it should be noted that, as far as consumers are concerned, they could lose either way. Subsidies will help enhance profitability of the manufacturer and may do nothing to reduce prices. The consumer may benefit through purchase of a crime-proofed item, but was the higher price paid for the item worth it? We see here the difficulty in translating the social costs of a product into something that the consumer will actually want to pay for. Thus, if the consumer does not see that the reduction in risk of theft is worth, say, an increase in price of the product of 5% of the cost of its production, the manufacturer is put in an extremely difficult position in marketing this crime-proofed product.

The third remedy to negative externalities is regulation. Depending on the product, government regulation may take on many different forms. For example, the history of regulation of auto safety in the U.S. reveals an incredibly complex web of government legislation, regulation through rules by government agencies, lobbying by consumers, manufacturers and insurance companies as well as lawsuits (Newman, 2004). By their action, governments may also directly affect markets. In the U.K., the extension of S17 of the 1998 Crime and Disorder Act, stating that all statutory agencies have a duty to reduce crime, could require that governments purchase only products secured by design. This would, presumably, shape the market so that goods with security commensurate with their CRAVED level would be favored.

It is reasonably well established that businesses are more adversely affected by fines and regulation than by subsidies. Usually regulation of some sort is a necessary part of a fines system, but both fines and subsidies do require some kind of monitoring, which is a cost that has to be taken into account. This is really a huge problem for it is clearly not in the economic interests of the company to reveal the criminogenic features of its products, particularly in a climate of regulation and fines. But it may be worthwhile for the company to do so in the case of subsidies if these are large enough. Thus, the transaction or monitoring costs of fines and regulation are generally much higher than the cost of monitoring subsidies.

Voluntary Codes

We are now in a position to consider how the idea of voluntary codes compares to the mandated remedies just described. We can consider this according to the criteria just expounded—the transaction or monitoring costs, the information requirements, and the problem of non-compliance. First, it is obvious that, if the entire system of codes is voluntary, there can be no monitoring costs to the government, though there may be some initial costs in providing the financial support or incentive to get the codes started in the first place. But once done, the government has no direct role in monitoring compliance. However, we say "direct role" because there may be an indirect role for government in helping provide information.

This leads to the second issue for voluntary codes: how to monitor compliance? Clearly, compliance cannot be monitored or established without the relevant data. Thus, a system or organization for the collection of information on compliance is necessary: this system would assemble comparative data on the rate of theft of particular products and product categories. Such a system would have to be independent of the manufacturer, since, as we have seen it is not in the manufacturer's interest to give up information concerning the criminogenic properties of its products. Therefore, a government agency would likely be a more independent collector of information than would the manufacturers themselves. However, it might be possible to construct an independent organization through a partnership between trade associations and government, along with other trade associations that may have a direct stake in the monitoring of criminogenic products, such as insurance companies.

Finally, we have seen that a voluntary code's greatest problem is that of non-compliance. At least with standards backed up by fines or regulation, there is a strong negative incentive for all to comply. But with a voluntary code, there is no obvious economic incentive to comply. Thus, there still must be some mechanism to establish compliance. If not, the freeloaders have a significant advantage. Apart from the necessity of data collection already noted, there is the clear necessity to create an incentive for manufacturers to comply. There are four possible ways to achieve this:

- The first is to create an economic necessity in the market that will induce manufacturers to design crime prevention into their products. The only mechanism available appears to be to educate consumers about criminogenic products so that they will begin to demand products that are resistant to crime, just as they now demand that products they buy should be designed to be safe (e.g., cars, child car seats, fireproofed clothing, toys etc.). One, perhaps unpopular way to advance customer education would be to publicize on an annual basis those products that are most at risk to be stolen. This has been done successfully with other products, especially cars.
- The second is to work with manufacturers so that they can develop ways to package the security features of their products along with other

features that they know customers want. This has been achieved with other products, including telephone services such as Caller ID and tamper-evident packaging. The latter was marketed promising not only security and safety but also quality (Clarke & Newman, op. cit.).

- The third way is for the government to pass laws that make it easier to bring class action suits for insecure products. While this is an inefficient way to achieve compliance because of the vagaries and complexities of civil law, there is a long history of successful suits against manufacturers of unsafe products, which suggests that this could be a powerful way to help create the incentive among all manufacturers to comply with a voluntary code. The success of such suits, however, will in the long run not depend on the laws passed by a government, but on the value that consumers place on secure products. Unless product security is valued, it is likely that class action suits will suffer from lack of public support.[16]
- The fourth way is to develop a partnership with insurance companies to ensure that they work to help implement the above three interventions and that their marketing practices do not work against them. The insurance industry has played a key role in the introduction of safety and security features of cars. It did this through powerful use of publicity and incentives to consumers with rate reductions for safety and security features (Newman, 2004). However, it can also counteract efforts by providing insurance to manufacturers who decide that it is cheaper to cover their liability through insurance rather than to the design or retro-design security into their products.

PART 2: ELEMENTS OF THE CODE

The information reviewed above leads to the conclusion that in formulating a set of codes, and the incentives for implementing them, we should be guided by the following principles:

1. Begin at an elementary level to set up a general framework for action.
2. Mindful of costs involved in setting up monitoring and regulation, begin with a voluntary code.
3. Work with the market itself to create incentives for code adoption.

In what follows, we outline the basic elements of a security code for the consumer electronic products identified earlier (those serving communication, entertainment and personal productivity), and limited to the prevention of theft of those products. For reasons stated earlier, other crimes have been excluded so as to simplify this task, since the components of risk analysis will vary according to crime type. For similar reasons, many product types, including electronic services conceived as products, have been excluded.

By "elementary" we mean to begin by constructing an easy to use checklist of items that define product risks of theft, along with a checklist of compensatory security features that might be added on to or designed into the product. The output of these checklists will be combined to determine whether or not it is likely to be stolen. We envisage a kind of "housekeeping seal of approval" endorsement, such as *Designed against Crime*, that could be applied to the product. This would be awarded to any product designated as low risk for theft. It would be promoted along with a logo and could be used in advertising of the product. After testing, this methodology could be adapted and extended to other types of products and types of crimes.

In this way, customers might begin to look for the *Designed against Crime* stamp of approval and might thus create a demand for manufacturers to design security into their products. Similarly, those products rated at highest risk could be publicized through the media, again allowing the market to apply its positive and/or negative incentives. The nature of the organizational capacity needed to administer this seal of approval raises some difficult issues and will be discussed after we have described the elements of the code. The three essential elements are: (1) assessing the inherent risk of theft; (2) rating the security features; and (3) determining whether the product has been "designed against crime."

The Checklist for Risk of Theft

Ideally, the risk of theft for any particular product ought to be measured on a routine basis, much as the Home Office now routinely measures the risk of theft for individual makes and models of cars. In theory, a combination of surveys, insurance data and retail shrinkage data could produce the needed risk scores, but this would be a laborious and difficult undertaking. We have recommended below that work should start along these lines, but to wait for its completion would delay the introduction of security code for manufactured products for some time, perhaps for many years. It would also mean that it would be impossible to security code new products until they had been on the market for at least a year or two. By this time, criminals might have enjoyed a substantial "crime harvest" (Ekblom, 1997). Accordingly, we have devised a methodology for security coding that relies on predicting the risk of theft for any product. Since this risk cannot be predicted with any degree of certainty, we have also provided for an informal review of a product's theft history on an annual basis and, if necessary, a recoding of the product. This review would be based on information supplied by the police and the insurance industry, a process that would eventually be replaced by the more formal calculation of risk discussed in the previous paragraph.

The methodology for predicting a new product's risk must be independent of technology, and must be capable of application to any electronic product as defined earlier. It also requires us to identify the essential components of risk in regard to theft of a product, and to identify the essential security features that may be expected typically to be employed to offset any security defects in the product design.

The first requirement for predicting a product's risk of theft is to assess its inherent attractiveness to thieves, i.e., what makes it "hot." We have chosen to define these elements using CRAVED, now a well-established means of identifying the extent to which products may be at risk to theft (Clarke, 1999). Goods that are hot are generally **C**oncealable, **R**emovable, **A**vailable, **V**aluable, **E**njoyable and **D**isposable. A brief description of these elements is as follows:

- *Available.* Availability is a necessary condition of being hot. This relationship can appear at the "macro" (or societal) level in mini theft waves resulting from the introduction of some new attractive product, such as the mobile phone, which quickly establishes its own illegal market. At the "meso" (or neighborhood) level, availability can show up in terms of the accessibility of hot products to thieves. For example, the fact that cars become at greater risk of theft as they age may be because they become increasingly likely to be owned by people living in poorer neighborhoods with less off-street parking and more offenders living nearby. At the "micro" (or situational) level, availability may show up in terms of the visibility of objects at the point of theft. This is why householders often conceal jewels and cash in the hope that they will not be found by burglars.

- *Valuable.* Thieves will generally choose the more expensive goods, particularly when they are stealing to sell. When stealing for their own use, other components of value become important. Thus, juvenile shoplifters may select goods that confer status among their peers, but which may not be expensive. Similarly, joyriders are more interested in a car's performance than its financial value. Two components of value, apart from monetary worth, merit separate treatment: the enjoyment of owning and using particular goods, and the ease or difficulty of selling them.

- *Enjoyable.* Hot products tend to be enjoyable things to own or consume, such as liquor, tobacco and CDs. Thus, residential burglars are more likely to take DVD players and televisions than equally valuable electronic goods, such as microwave ovens or food processors. This may reflect the pleasure-loving lifestyle of many thieves (and their customers). Interviews with street robbers have found that the majority of offenders say they rob for money and that they spent the money on expensive clothes, other luxuries and cannabis.

- *Disposable.* It may be obvious that the thief will tend to select things that will be easy to sell, but its importance for explaining crime has been neglected. Only recently has systematic research begun on the intimate relationship between hot products (including electronic products) and theft markets (Sutton, 1998; Kock et al., 1996).

- *Removable.* Products that are easily moved are more likely to be stolen, a fact well understood in security practice. How easily security can be defeated depends on the circumstances of theft. This point is substantiated by American data showing differences in what is stolen from supermarkets by shoplifters and burglars. Both groups target cigarettes, liquor, medicines and beauty aids, but these are taken in much larger quantities by the burglars (Food Marketing Institute, 1997).

- *Concealable.* Items are less likely to be stolen: (1) that cannot be concealed on the person, (2) that are difficult to hide afterwards, and (3) that are easy to identify. This explains why we write our names in books and why cars must be registered and licensed. It also helps explain why car thieves will not generally steal a Rolls Royce for their own use, but will steal instead a less valuable car that merges into the surroundings. The same principle also helps explain why cars stolen in the United States for export to Mexico are mainly models that are also sold there legitimately. Consequently, stolen cars do not stick out like sore thumbs (Fields et al., 1992).

These components of CRAVED form the basis of the checklist used to establish a product's attractiveness for theft. We have made only two modifications for this purpose, the first of which is to specify a little more the measurable elements of each component that would be relevant to the range of electronic products selected. Second, we have included items related to marketing or sales strategies that may aggravate or increase the "hotness" of products. For example, advertising targeted at young males may establish a heavy demand for the product among that sector of the population most likely to be involved in theft, and aggressive promotion of the enjoyable features of owning or using the product could also contribute to its attractiveness for theft. We should have liked to include items measuring the social cost of a product, but this will only be possible once detailed data are available about the actual risks of theft for particular products. We have noted above that such data exists for cars, but collecting similar data for other products would be a formidable undertaking. Once available, however, Field's methodology used for calculating the costs of car theft in the United States could probably be adapted for other kinds of products (Field, 1993).

Checklist One displays the checklist for determining a product's inherent risk of theft. We should note that the elements rated and the scores assigned to each are subject to change following the piloting of the two checklists, which we recommend is undertaken as one of the immediate steps needed to implement the code.

The Checklist for Product Security

The second requirement for establishing a product's risk of theft is to rate the extent to which security features have been adopted to offset the product's intrinsic attractiveness for theft. For example, the small size of a personal CD player, which makes it concealable, cannot be altered without destroying the entire concept of this product. This may be neutralized by an added security feature, such as packaging, which would make the shape and size of the product more difficult to conceal on the person. It may also be possible to reduce the "stealability" of the product through customer education concerning its use, or through training of retailers in adopting customer education in their sales approach.

Checklist One: How "Hot" Is the Product?

	Items	Item Score
	Check one	
CONCEALABLE	❒ on person (score 2)	
	❒ in bag (score 1)	
	Check one	
REMOVABLE	❒ can be carried in one hand (score 2)	
	❒ can be carried with two hands (score 1)	
	Score 1 for each:	
AVAILABLE	❒ used outside the home	
	❒ commonly left in parked cars	
	❒ marketed to young males	
	❒ minimal search time for thief to locate product	
	Score 1 for each:	
VALUABLE	❒ costs at least one day's wages	
	❒ provides access to phone service	
	❒ provides access to Internet	
	❒ provides access to credit	
	Score 1 for each:	
ENJOYABLE	❒ entertaining	
	❒ addictive	
	❒ fashionable	
	❒ luxury item	
	❒ status item	
	❒ aggressive advertising emphasizing these themes	
	Score 1 for each:	
DISPOSABLE	❒ widely in demand	
	❒ value easily assessed	
	❒ street price less than 50% of one day's wages	
	TOTAL SCORE	

Checklist Two displays a variety of security features that might be included in a product's design or marketing so as to reduce the likelihood of theft, or, in the case of the replacement guarantee, at least offset the cost for the consumer. This guarantee might in fact become a source of profit for manufacturers, especially if it were bundled with after-sales service agreements. Product identification features such as RFIDs may also help in recovery of the item should it be stolen.

Determining *Designed against Crime* Status

The scores from the two checklists provide the basis for determining *Designed against Crime* status for any particular product. In principle, the greater the inherent risk of theft, the greater must be its security for a product to qualify for *Designed against Crime* status. Thus, if the "hottest" products, with the highest scores on Checklist One, are to qualify, they must also obtain high scores on Checklist Two for the security built into their design or marketing. Conversely, the products that are least attractive to thieves would require lower levels of security to qualify.

In order to establish the range of scores qualifying a product for *Designed against Crime* status, a sample of electronic products (say 60) must first be rated using the checklists in order to establish a range of scores. On the basis of this exercise, it should be possible to sort each product into one of three categories of inherent risk (low, medium and high) and into three categories of security (again, low, medium and high). While in principle not difficult, this grading is likely to require some further adjustment of the scores attached to individual elements of the checklists. The final step will then be to compare the two sets of scores. Products with a "high" risk of inherent theft must also fall into the group with "high" security to be designated as *Designed against Crime*. Those with "medium" risk need only be "medium" on the security scale to qualify, and those with "low" risk need only be "low" security. As experience with use of the checklists accumulates, it is likely that the scoring of the items and the ranges of scores (high, medium and low) for each checklist would need to be adjusted and refined.

Procedure for award of *Designed against Crime* status:

1. Manufacturer decides to apply for *Designed against Crime* for a particular product.

2. Undertakes an analysis of inherent risk of theft with Checklist One.

Checklist Two: Product Security Features

Security feature	Score
❐ Customer education designed into marketing (e.g., security instructions included in package) (score 1)	
Replacement guarantee to consumer if product stolen. Check one:	
❐ Within 90 days (score 1)	
❐ Within 1 year (score 2)	
❐ Life of product (score 3)	
❐ Tracking technology such as RFIDs to make recovery of item easier if product is stolen, particularly during the journey from manufacturer to consumer (score 3)	
❐ Customer education to minimize risk of theft of product included in retailer training (score 1)	
❐ Valid means of unique identification of product (e.g., source tagging) (score 3)	
❐ Valid means of tracking ownership of product through life cycle (e.g., chipping) (score 3)	
❐ Technology designed to delay or defeat attempted theft of item (e.g., packaging) (score 3)	
❐ Technology to negate the financial value of the item if stolen (e.g., PIN) (score 3)	
Cost of inclusion of security features has been:	
❐ 10% or more production cost (score 2)	
❐ Up to 10% of the production cost (score 1)	
❐ Zero cost (score 0)	
Cost of security feature included in product has been:	
❐ Absorbed by manufacturer (score 2)	
❐ Shared with retailer (score 1)	
❐ Shared with customer (score 0)	
❐ Passed on to customer (subtract 1)	
Product has been field-tested for theft*	
❐ Yes (score 1)	
❐ No (score 0)	
TOTAL SCORE	

*Field-testing consists of market research into the product's perceived attractiveness to thieves.

3. Completes assessment of security measures using Checklist Two, and makes any necessary improvements in security.
4. Submits plan to responsible trade association (see below).
5. Trade association grants *Designed against Crime* seal of approval (or not).
6. Trade association reviews theft history on annual basis to determine any needed improvement in security to retain *Designed against Crime* status.

Administration of the Code

Organizational Need

An organization must be established to administer this voluntary code and supervise the collection of data needed for the code to work in a reliable way. The obvious question is whether government should be part of this organization. Our preference is that, unless extensive groundwork were done to establish a partnership between business and government to collect the appropriate data, there would be great difficulty in a government department collecting data that are sensitive to the operations of business, and that, used inappropriately, could even negatively impact business activities. Therefore, we think that the industry itself through its trade association, perhaps in partnership with the insurance industry, should form an organization to both fund and administer the code. This solution preserves an open market approach to voluntary codes, but at the same time creates competition between two parties with different interests in the market. Together, this should produce reasonably unbiased and reliable collection of information.

Government Participation

The initial establishment of the organization to administer the code and collect relevant data would certainly be helped by government working in partnership with the industry trade association to convene meetings among the interested parties and to provide the necessary financial support. The Home Office also has extensive crime prevention expertise that could be of great assistance in the development of the code.

IMPLEMENTING THE CODE: NEXT STEPS

Implementing the code will involve a number of steps, as laid out below. This assumes that a designated government unit will play a substantial part in bringing together the various partners (other government departments, large electronics manufacturers, electronics trade associations, retailers associations, insurers, police, etc.) whose participation will be needed to introduce the code. Only a government unit would have the necessary authority and independence to perform this role, but it must be provided with the research capacity needed in at least the early stages of the code's development. As explained above, we also envisage that the electronics industry's trade association(s) will assume an increasingly large part in administering the code. However, if the code is to serve as prototype for codes that might be applied to other products and industries, the said government unit will, for the foreseeable future, have to take the lead in such an extension of the work.

Immediate Steps

1. Take soundings from the electronics industry on the plan for the proposed code and revise and extend as necessary. This process has been started by the Jill Dando Institute of Crime Science, which has circulated this paper for comment to the British Radio and Electronic Equipment Manufacturers Association. Together with the Panel Secretariat, the Institute will also visit the Consumers Association to obtain that body's views on the proposed code.

2. Once these preliminary soundings have been taken and any necessary modifications have been made to the plan, a series of meetings should be convened with manufacturers, trade associations, and other interested parties to obtain wider feedback on the feasibility of the code and on practical ways to apply it. It will be particularly important to obtain the views of retailers on whether security of products can be used as an effective selling point and whether the benefits of a *Designed against Crime* stamp of approval will be easy to explain to the customer.

3. Obtain legal advice on the proposed code in the light of experience gained from similar "secured by design" schemes in parking lots and housing.

4. Alert the "chipping of goods" initiative.

5. Rate a sample of electronic products (say 60) using the checklists in order to establish the range of scores determining bands of low, medium and high for both risks and security.

Medium Term

1. Obtain economic advice on the conceptual framework, especially on the problem of estimating the social costs of products.
2. Consider the obvious parallels between product security and product safety. A project should be initiated to examine these similarities in regard to consumer demands, the generation and control of externalities, and the success or lack thereof in constructing codes of performance for products, and of attempts to obtain compliance.
3. Initiate a project to review the externalities generated by the Internet and their relationship to "hot services," and assess the viability of extending the codes to services.

Longer Term

1. Extend the code to electronic business products.
2. Extend the code and incentive plan to other consumer product types.
3. Consider extension of the code to other crimes besides theft (though this will entail considerable development work).
4. Encourage introduction of new technology to track and verify ownership of products through their life cycle.
5. Explore options for standardized registration of property.

Continuing

1. Educate business and industry concerning the need for security coding and the benefits of adopting such codes.
2. Educate the public concerning the benefits of security coding and the part it can play in demanding secure products.
3. Initiate work on the collection of data on actual product risks, and gain the cooperation of relevant bodies, such as insurance and retailing firms.

4. Set up a permanent, independent organization for the review of code compliance, collection of data on product theft, and return of information to the industry and the public.

CONCLUSIONS

When we began this task, the conceptual and practical problems of establishing a security code, even for a limited range of electronic products, seemed overwhelming. Several months later, having completed the work, we feel more sanguine. Provided that one begins in the simplest manner possible, yet on the basis of a solid conceptual footing, we think that such a code could be brought into force and that the plan we have outlined could serve as the basis for the code. At the same time, we concede that the plan is ambitious and that it will require considerable determination on the part of the industry and government if it is to be successfully implemented. It will also require careful planning and monitoring of the sequential steps needed to bring the code into force. Finally, it will require a long and sustained investment of resources in an enterprise that may seem quixotic to many people.

Having said all this, we believe that the expenditure of this energy and resources is justified by the potential benefits to society. It is safe to predict that technology will continue to deliver a cornucopia of new products and services, many of which will be electronic, that will be eagerly sought by the consumer. Since these will never be delivered at prices that everyone can afford, there will be no end to their theft and criminal exploitation. The challenge to those of us faced with this problem is to outsmart the offender, perhaps by harnessing the technology that is helping to create the problem. Security coding of products offers a promising route to that goal.

Address for Correspondence: Ronald Clarke, School of Criminal Justice, Rutgers University, 123 Washington Street, Newark, NJ 07102 (email: rvgclarke@aol.com).

Acknowledgments: This is a slightly modified version of a paper that the authors produced for The Foresight Crime Prevention Panel on behalf of the Jill Dando Institute of Crime Science, University College London. We are grateful to the

Jill Dando Institute and the Department of Trade and Industry for permission to reproduce the paper.

NOTES

[1]The Jill Dando Institute for Crime Science, University College London, is participating in an EU-funded project, led by Ernesto Savona of the Catholic University in Milan, to explore the feasibility of the plan laid out here for security coding of electronic products.

[2]The design of mobile phones clearly contributed to their theft in the United Kingdom (Harrington & Mayhew, 2002) and to their "cloning" and theft of cell phone service in the United States (Clarke et al., 2001)

[3]There is no research indicating whether there is variation in the theft of electronic products across brand names within a product category. This type of variation is, however, difficult to predict, depending on such matters as fashion, consumer preferences, market saturation, and product promotion. This does not mean that risk of theft could not be predicted for various brands. The way in which it could be done would be to collect information on a regular basis concerning theft of a variety of products, and construct prediction tables based upon those data. A systematic data collection so that prediction of initial risk can be compared with data on actual risk we consider to be an essential part of the implementation and maintenance of any crime-proofing code. It is not clear whether product design could reduce theft that resulted from the variations in distribution of risk, but most certainly the security response to risk should take this factor into account.

[4]Similarly, products may be stolen at the point of home delivery, as a result of catalogue or etailing sales, although there are limited data as to the prevalence of this type of theft (Newman & Clarke, 2003).

[5](Adams & Hartley, 2000). The introduction of chipping and advanced source tagging promises to make possible even greater tracking and identification of ownership of products as they move through their life cycle from manufacturer to customer, and even sequential customers. See for an example: http://www.smartwater.com/full_instant.htm and http://www.applied-holographics.com. And for a taste of RFIDs of the future see: "Where's the smart money?" *The Economist*, Feb. 7, 2002, p. 69.

[6]We qualify this observation because it is conceivable that hi-tech Smart Cards with far more technological features built into their design, could become devices similar to mobile phones and other hand held devices that people own.

[7]In fact, there could be general design guidelines that took account of safety factors in regard to appliances of any kind, such as avoiding sharp corners on any appliance or device, and reducing the overall weight of such objects. Reducing the weight, however, would also make the object more stealable. Clearly, there is an important interface between product design for safety and product design for security that has not to date been examined.

[8]Wall and Davis (2001) argue that the original acronym CRAVED used by Clarke to identify the criminogenic properties of products was not applicable to electronic

services and that a better substitute would be EVADED: **E**nduring (once appropriated can continue to be used), **V**aluable, **A**vailable, **D**istributable (can the thief distribute the service to others), **E**asy to use, and **D**esirable (may be desirable to a thief without the service having any monetary value). EVADED does not quite capture the features of hot services, if we think of them as part of an information system. Nor does EVADED clearly distinguish between the environment within which the service is provided—i.e., the information system, and the content of that information system, that is information. If we think of information as a product in and of itself, all the attributes of CRAVED apply incredibly well and therefore apply to that part of services that is composed of information. However, the other part of service is the information system, which is an environment that produces "hot situations" or criminogenic environments in which crime is made much easier to commit. These are described by the acronym SCAREM: **S**tealth, **C**hallenge, **A**nonymity, **R**econnaissance, **E**scape, and **M**ultiplicity (Newman & Clarke 2003).

[9]"Lunchtime @ kitchen-fridge," *Computer shopper* (2002). March. p.32. LG Electronics announced the release of its Internet linked refrigerator (www.lge.com).

[10]The CEO of Phillips in a speech on January 9, 2002 anticipated an "intelligent home environment that is sensitive, personalized, adaptive, anticipatory and responsive to people." Such an environment could not be established or maintained without extensive personal service oriented products, most likely administered and monitored through Internet technology: http://www.news.philips.com.

[11]Neither of the present authors has economic training and the discussion of externalities in this paper draws heavily on Stiglitz (1998).

[12]Crime is perhaps an artificial "common resource," having been created by laws. That is, without criminal laws there could be no crime, although there would be behavior that looked like what we call crime today.

[13]In economics this manner of dealing with externalities is known as the Coase Theorem (Coase, 1960).

[14]However, in the case of "hot services" that are offered on the Internet, externalities of a special sort could apply. For example, economists have observed that a special type of positive externality occurs on the Internet, where the value of a product/service increases exponentially because of the large addition of users on the Internet, thus creating a much larger possible market for these products and thus making them more valuable. The value added to such product/services is not a result of the company that produced them, but a result of additional users joining the network. That is, the producer/manufacturer of that service did not pay for the value received. (See: for example, John S. Irons. Network Externalities and the "New Economy." *Economics* (http://economics.about.com/library/weekly/aa030198.htm). If we think of a hot service as, say, a fraudulent web site selling bogus investment products, the value of that web site and the products sold is increased by the easy access to the site by larger numbers of Internet consumers. Rights and ownership of certain aspects of the "common property" of the Internet therefore could be at issue. This problem, however, takes us beyond the topic of our paper, although it does highlight the enormous

complexity that would be involved if we tried, in the present paper, to incorporate services as products into the codes/incentives framework.

[15]This is a highly simplified description of a system of fines. Since our overall approach is to consider voluntary codes, we do not give close attention to fines. However, it should be noted that there are a wide variety of fine schedules available and in use, such as non-linear fine schedules where fines remain very low up to a certain point, then jump to a very high amount.

[16]That this approach can be effective is clearly evident from other product categories, such as handguns in the United States. Taking their lead from the lawsuits brought against tobacco companies, state governments and private interest groups joined together to bring suit against handgun manufacturers. While the final outcome of these suits remains in question, there is little doubt that their mere threat has helped bring about the marketing of more secure handguns (Clarke & Newman, op. cit.).

REFERENCES

Adams, C., & Hartley, R. (2000). *The chipping of goods initiative. Property crime reduction through the use of electronic tagging systems. A strategic plan.* London: Home Office Police Scientific Development Branch.

Anderson, R. J. (2001). *Security engineering: A guide to building dependable distributed systems.* New York: Wiley.

Clarke, R. V. (1999). *Hot products. Understanding, anticipating and reducing the demand for stolen goods.* Police Research Series, Paper 112. London: Home Office.

Clarke, R. V., & Harris, P. M. (1992). A rational choice perspective on the targets of auto theft. *Criminal Behaviour and Mental Health, 2,* 25–42.

Clarke, R. V., Kemper, R., & Wyckoff, L. (2001). Controlling cell phone fraud in the US: Lessons for the UK "Foresight" Prevention Initiative. *Security Journal, 14,* 7–22.

Clarke, R. V., & Newman, G. (2005). Modifying criminogenic products: What role for government? This volume.

Coase, R. H. (1960). The problem of social cost. *Journal of Law and Economics, 3,* 1–44.

Ekblom, P. (1997). Gearing up against crime: A dynamic framework to help designers keep up with the adaptive criminal in a changing world. *International Journal of Risk, Security and Crime Prevention, 2,* 249–265.

Field, S. (1993). Crime prevention and the costs of auto theft: An economic analysis. In R. V. Clarke (Ed.), *Crime prevention studies* (Vol. 1). Monsey, NY: Criminal Justice Press.

Field, S., Clarke, R. V., & Harris, P. (1992). The Mexican vehicle market and auto theft in border areas of the United States. *Security Journal, 2,* 205–210.

Food Marketing Institute. (1997). *Security and Loss Prevention Issues Survey.* Washington, DC: Food Marketing Institute.

Hardie, J., & Hobbs, B. (2001). *Companies against crime.* Draft paper for IPPR Criminal Justice Forum (Draft 10-01-02). A slightly revised version is included in this volume.

Harrington, V., & Mayhew, P. (2002). *Mobile phone theft.* Home Office Research Study, No. 235. London: Home Office.

Kock, E., Kemp, T., & Rix, B. (1996). *Disrupting the distribution of stolen electrical goods.* Crime Detection and Prevention Series, Paper 69. Police Research Group. London: Home Office.

Newman, G. (2004). Car safety and car security: An historical analysis. In M. Maxfield & R. Clarke (Eds.), *Crime Prevention Studies, 17,* 213–244.

Newman, G., & Clarke, R. V. (2003). *Super highway robbery: Preventing e-commerce crime.* London: Willan.

Stiglitz, J. E. (1988). *Economics of the public sector* (2nd ed.). New York: W. W. Norton.

Sutton, M. (1998). *Handling stolen goods and theft: A market reduction approach.* Home Office Research Study, No. 178. London: Home Office.

Wall, D., & Davis, R. (2001) *Turning the corner,* CD annex. London: Department of Trade and Industry.